GARDEN BIRDS & WILDLIFE

BTO
Looking out for birds

'**Power-packed with fascinating facts**'
TONY SOPER

GARDEN BIRDS & WILDLIFE

Mike Toms and Paul Sterry

FOREWORD BY CHRIS BEARDSHAW

Written by Mike Toms and Paul Sterry

© AA Media Limited 2011
© Text BTO and Paul Sterry
© Digital imaging and manipulation Paul Sterry

Produced for AA Publishing by Paul Sterry and Shane O'Dwyer
Designer: Shane O'Dwyer
DTP Management: David Price-Goodfellow, D & N Publishing,
   Baydon, Wiltshire

Production at AA Publishing: Stephanie Allen

Printed in China by C & C Offset Printing Co. Ltd

A CIP catalogue record for this book is available from the British Library.

ISBNs 978 0 7495 7149 8 and 978 0 7495 7162 7 (SS)

Published by AA Publishing (a trading name of AA Media Limited,
whose registered office is Fanum House, Basing View, Basingstoke
RG21 4EA; registered number 06112600).

A04633

The Publishers would like to thank Paul Sterry, Shane O'Dwyer, Mike Toms
and David Price-Goodfellow for their devotion to this project. This book
was only possible because of them.

The authors would like to express their gratitude to everyone involved with
the creation of this book. But particular praise is due to Shane O'Dwyer
for his inspirational design and attention to detail.

CONTENTS

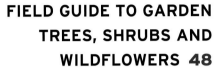

FOREWORD

As a young boy I recall being despatched to bed after a long day of summer play in the garden and farmland that surrounded my home. Resentful at my enjoyment and exploration being curtailed, I distinctly recall the amber sunlight of evening spilling though the gap in heavy curtains and, as I lay on cool sheets, being lulled to sleep by the harmonious song of our resident Blackbird, perched on the cottage chimney pot. As if celebrating the day past and announcing the onset of dusk, hearing the song of this most optimistic and delightful bird still instantly transports me to my childhood, while the echoed calls of the Blackbirds that share my garden today mark the point at which now my children depart for their beds.

I am sure we all retain fond memories of close encounters with nature, from the inquisitive and cheeky Robin that utilises a spade handle as a perch, to the fleeting glances of Pipistrelle bats as they perform their acrobatic aerial displays against a harvest moon. Such interactions and observations provide a unique dimension to life outdoors, and to be honest it hardly matters how rare and unusual an organism might be: the critical element is the recognition that wherever we stand the beauty of nature is never far away. And no matter how familiar, there are some organisms that never fail to impress: millipedes silently scurrying with military precision over leaf litter in woodland have long entertained me,

while the ghostlike sweeps of the Barn Owl that occasionally graces our garden is guaranteed to stimulate a tremendous sense of privilege.

While nature enriches our lives it is also important to remember the role we all play in providing a resource for wildlife. Whether creating a garden specifically for wildlife or simply considering adjusting the maintenance of an existing plot, there is an increasing need to adopt more holistic and sympathetic practices to ensure wildlife is offered a diverse foundation. Key to this is the identification and understanding of those animals and plants with which we share our lives. And it is here that *BTO Garden Birds and Wildlife* excels, providing an intimate and detailed observation of the key wildlife personalities and characters in a format that allows easy access to both bullet-point details and up-to-date scientific data. Armed with this book and a willingness to engage, observe and entice nature, a new and enlightening beauty will be revealed. The fact-filled pages of *BTO Garden Birds and Wildlife* are also a testament to the sterling research undertaken by the BTO.

Ultimately for me there is no doubt that the sights, smells and sounds of nature in all her guises provide essential animation, emotional catalyst and unrivalled inspiration to anyone willing to take the time to experience them. I would even go as far to argue that a garden or landscape without wildlife is a place without a soul.

CHRIS BEARDSHAW

Starlings are regular visitors to many gardens. With their bold and pugnacious behaviour they are birds with real 'attitude'!

A well-stocked feeder with a number of different feeding ports will attract a range of species, in this case six House Sparrows, a Greenfinch and a female Chaffinch.

INTRODUCTION

GARDENS ARE IMPORTANT for many bird species, some individuals being resident year-round, others visiting when food is in short supply elsewhere. Establishing how and when birds use gardens can make an important contribution to our understanding of bird populations and how they alter in response to changes in the wider countryside. With increasing pressures on land for housing and food production, this type of information is needed to underpin conservation strategy and to secure bird populations for future generations. Much of the information on how and when birds use gardens comes from amateur naturalists who contribute to national surveys, particularly the BTO Garden BirdWatch, by keeping simple records throughout the year of the birds that use their gardens. It is fitting that their sterling work provides the basis for the facts and figures that adorn this book. In compiling *Garden Birds and Wildlife*, the intention has been to help you, the reader, develop your interest in birds and gain some insight into their daily lives. It is also hoped that it will guide you from armchair birdwatcher to citizen scientist, so that you can contribute to the next generation of garden records!

Birds are not the only organisms to use gardens and a typical garden may support a wide range of different species, from tiny microbes to medium-sized mammals. While birds may be the most accessible of the groups to use your garden (and probably the reason that you picked up this book), it is likely that you will also have an interest in some of the other species that are present in your garden. For this reason, we have also included sections on other garden wildlife, helping you to identify many of the species that might be encountered throughout the course of a year. It is also likely that you might wish to increase the attractiveness of your garden to wildlife and so we have included sections on wildlife-friendly gardening, recommending suitable plants to attract insects, and berry-producing shrubs for small mammals and birds.

Finally, and perhaps most importantly, gardens provide a unique opportunity for people to engage with wild creatures at first hand. This contact is particularly important for those who do not have access to the wider countryside, perhaps because they live in a city or suffer from poor mobility. With an ever-increasing proportion of the population living within urbanised landscapes, gardens are likely to take on an increasingly important role in providing contact with the natural world. As such, it is essential that we seek to understand their role in supporting different species of plants and animals and to ensure that our gardens are as beneficial to wildlife as possible.

DID YOU KNOW ?

Fifty years ago, anyone interested in ornithology would have called himself or herself a 'birdwatcher', without a second thought. But although we are all *bird watchers* at heart some people categorise themselves, or others, as 'birders' or 'twitchers'. Are these real distinctions? The term 'birder' is a relatively recent one and is someone who tends to be focused on finding and studying birds, often concentrating on a local patch. By comparison, traditional 'birdwatchers' tend to spend less time in the field than 'birders' and often visit nature reserves. A 'twitcher' is someone who races around the country after rare birds – found by other people – so they can tick them off their list; twitching is an extremely competitive pursuit. The truth is there is an element of the 'birdwatcher', 'birder' and 'twitcher' in most ornithologists and regardless how we think of ourselves, we are all united by a passion for birds.

MIKE TOMS and
PAUL STERRY

HOW TO USE THIS BOOK

GARDEN BIRDS AND WILDLIFE serves a number of functions. Primarily it is a resource of accurate and contemporary information about birds in the context of our gardens. It allows the reader to gain informed insight into the lives, trials and tribulations of the nation's most familiar species. Although this book's role is not that of a field guide, the basics of garden bird identification have been catered for too, and recognition of common native garden plants and animals has also been covered. Lastly, *Garden Birds and Wildlife* is also a visual celebration of garden birds, allowing the reader to appreciate the true beauty of the birds with which we share our lives.

Undeniably the most accurate information about British birds - up-to-date facts, figures and details about food, nesting and lifespan; condensed from the British Trust for Ornithology's online database *BTO BirdFacts*.

BTO FACT FILE

Stunning colour photographs illustrate a wide range of plumages, poses and behaviours for every species found in the garden. Some of the images were taken especially for *Garden Birds and Wildlife*.

PHOTOGRAPHS

Using information collected, collated and interpreted by several generations of BTO researchers and volunteers, and written in an easily accessible style, the main text provides the reader with a wealth of information about garden birds.

MAIN TEXT

Unusual and unexpected facts about British garden birds are presented in 'IT'S A FACT' boxes that are distributed throughout the book; some relate to the species in question's plumage, others to behaviour or distribution.

IT'S A FACT...

BTO BIRDFACTS

ROBIN
Erithacus rubecula

FAMILY
Turdidae
LENGTH
14cm
WINGSPAN
21cm
WEIGHT
18g
HABITAT
Woodland edge, scrub and gardens.
FOOD
Insects, especially beetles, spiders, together with fruit and seeds.
NEST
A bulky structure with leaves for the base on which is woven a cup of moss, grass and more leaves. The cup is lined with finer material, including hair and, occasionally, feathers.
EGGS
4–5 eggs, white or faintly bluish white with sandy red freckles and small blotches, which may sometimes make the egg appear reddish.
EGG SIZE
19.9 × 15.4mm
INCUBATION
14-16 days. 2-3 broods on average.
BREEDING SEASON
March to July
POPULATION SIZE
5.5 million pairs.
TYPICAL LIFESPAN
2 years
MAX. RECORDED AGE
8 years 4 months

Voted the Nation's favourite bird on more than one occasion, the Robin is also one of the most widely distributed of our birds. Because of their approachability and choice of habitats, Robins have been the subject of many studies, including that carried out by David Lack and famously published in 'The Life of the Robin'.

CATHOLIC AND CONFIDING
Robins are catholic in their choice of food items, although most of the diet consists of invertebrates taken from the surface of the ground. Fruit is also important, especially in the autumn and winter. They have been recorded feeding at hanging peanut feeders, and even hovering at hanging fat blocks, but they are happier taking mealworms, peanut cake, fat and finely grated cheese from bird tables or off the ground. Their confiding nature is further demonstrated by the way in which they will often associate with a gardener digging over the soil. This behaviour is probably an extension of the habit that Robins have of taking insects from ground disturbed by foraging moles. Robins usually use one of two main strategies for finding food. Individuals will sometimes sit on a low branch, scanning the ground beneath them. Every now and then they will spot a potential prey item, fly down to grab it, and then return to the perch. The other strategy involves a bit more legwork, with the individual hopping across the ground actively looking for food.

Juvenile birds, recognised by their spotted plumage, are dependent on the parents for food for a short while after they fledge.

YEAR ROUND SONG

Most male Robins defend territories throughout the year, which is why their plaintive song can be heard in every month. The territory structure adopted by Robins is all-important and only breaks down during periods of severe winter weather or when the birds are skulking away during their annual moult. Many females also establish winter territories of their own, often close to where they will breed the following year. Other females appear to move farther away from their breeding territories, sometimes setting up winter territories in habitats that could not support a nesting pair but which can support an individual Robin through the winter. Towards the end of the winter, these females return to their original territory and partner from the previous year.

MOVING AROUND EUROPE
A small number of our Robins appear to spend the winter on the Continent, joining other Robins that pass through southeastern Britain in the autumn en route south. These passage migrants come from Scandinavia, Germany and the Baltic states and are heading towards wintering grounds in Spain, Portugal and North Africa. Studies have shown that many return to the same winter territory each year, suggesting that these birds have distinct summer and winter territories, many hundreds of kilometres apart, to which they faithfully return each year.

ABLE TO BOUNCE BACK
As BTO Garden BirdWatch results demonstrate, Robin populations have been remarkably stable in recent years. Although they can be knocked back by very bad winters, the ability to produce three or even five broods in a year means that populations recover quickly from overwinter losses. In mild winters, breeding may begin in January and pairs may overlap nesting attempts in order to squeeze in several broods.

WHO'S THE ROBIN IN CHARGE?

Robins seem to take territorial behaviour to an extreme, sometimes attacking other species by mistake. The Robin's red breast is displayed prominently during territorial disputes. Initially, two birds disputing ownership of a territory will sing at each other, the individual owning the territory attempting to take a perch above the intruder. This enables the owner to show off his red breast to maximum effect by adopting a horizontal position and, in most cases, this is usually enough to settle the dispute. If the territory owner finds himself below the intruder then he will adopt a different posture, throwing his head back. Again this emphasises the red breast. Sometimes this display does not do the trick and the dispute escalates to an all-out attack. Such fights can be ferocious and injury or even death may result. Given the risks involved, it is hardly surprising that the birds first attempt to settle the dispute through display rather than all-out aggression.

IT'S A FACT...
The importance of the red breast for territorial display explains why young Robins are not born with this feature but acquire it during their first moult.

THE NEST

Robins will take readily to open-fronted nestboxes, so long as these are positioned among some thick cover, such as ivy growing on a wall, where they are afforded some protection from the unwelcome attentions of predatory cats. They will also use garden sheds and greenhouses where considerate owners leave the door ajar.

Robins are colourful even in winter.

STATUS

Population largely stable, hence Green listing.

SEASONAL TRENDS

BTO Garden BirdWatch reporting rate (%) throughout the months of the year

---

ILLUSTRATIONS

Garden Birds and Wildlife's lavish illustrations have been created by some of Britain and Ireland's finest artists. Some of the illustrations portray interesting behaviour while others focus on accurate detail for identification purposes.

FEATURE BOXES

Some aspects of the behaviour, plumage or occurrence of garden birds are more significant than others and subjects of particular interest to the reader are identified and highlighted in feature boxes.

STATUS and SEASONAL TRENDS

STATUS - the text provides a summary of the species' conservation status while a map shows its range, seasonal occurrence being represented by different colours: ■ = PRESENT YEAR ROUND; ■ = SUMMER VISITOR; ■ = WINTER VISITOR.

SEASONAL TRENDS - BTO Garden BirdWatch reporting rate; the graph line's colour reflects the species' conservation status.

THE BRITISH TRUST FOR ORNITHOLOGY – BIRDWATCHING AND SCIENCE IN PARTNERSHIP

The British Trust for Ornithology fulfils a vital role in the contemporary birding scene. It empowers a small army of volunteers and provides unrivalled analysis of the status of, and trends in, British bird populations.

ORIGINS OF THE BTO

From the outset, the British Trust for Ornithology comprised professional (paid) ornithologists working in partnership with interested amateur (volunteer) birdwatchers. The concept behind this approach was that the Trust would act as the nucleus, stimulating and co-ordinating research into populations of wild birds and their changing fortunes. The central core of this work would be periodic surveys, looking at specific issues, coupled with regular, often annual, monitoring of bird populations within different habitats. Such ambitious plans required the support and involvement of a large number of field ornithologists (birdwatchers), operating across the country in a systematic manner. Since its establishment, this partnership between birdwatchers and researchers has proved to be incredibly successful and the BTO and its volunteers are now the main source of ornithological information within the United Kingdom. When the Government and conservation bodies draw up conservation policies it is invariably the records collected by the BTO that are used to determine what should be done. It is because of BTO volunteers that we know where our Swallows spend the winter, why our Song Thrushes are in decline and how our bird populations are responding to global climate change. Key to this success has been the independence of the BTO; as an organisation it remains impartial, never lobbying on a particular issue or campaigning for a conservation cause. By remaining independent and impartial the information generated by the BTO can be regarded as being unbiased, allowing it to be used by Government and others to make accurate assessments of what is happening to our bird populations.

THE VALUE OF AMATEUR NATURALISTS

There is a long history in this country of amateur naturalists contributing to our understanding of the wild creatures and plants alongside which we live. Until relatively recently, the collection of records and observations was largely spontaneous in nature and unorganised in structure, with efforts often targeted towards individual species rather than populations, plant and animal communities, or ecosystems. The establishment of organisations like the Royal Society for the Protection of Birds helped to focus attention on conservation issues surrounding the changing fortunes of British and Irish birds but a systematic understanding of what was happening (and why) was missing. Against this backdrop, the fledgling British Trust for Ornithology (BTO) emerged in 1933.

ANNUAL MONITORING – PULLING IT ALL TOGETHER

One of the keys to the success of the BTO approach has been its ability to draw upon information collected through a series of different surveys and schemes. This approach, known as Integrated Population Monitoring (IPM), enables the BTO to collate information on survival rates, with that on population size and productivity. By doing so, it is possible to tease out the underlying causes behind observed population change. For example, knowing that the productivity of a species (the number of chicks fledging) is stable, while adult survival and population size are declining, would suggest that the cause of an observed population decline is due to something affecting adult birds. Future work can then be targeted towards this part of the life cycle in order to determine the ultimate cause of decline.

The BTO's headquarters are in the tranquil setting of The Nunnery at Thetford in Norfolk.

ABOVE: **BTO scientists removing a Sedge Warbler from a mist net, prior to it being ringed.**

LEFT: **Fieldwork being undertaken as part of the BTO's Common Bird Census.**

THE NATIONAL RINGING SCHEME

Take a look at your garden on a winter morning and you might be greeted by the sight of Blackbirds, Siskins and Chaffinches, all gathered together to feed on the treats that you have provided. What you may not realise is that some of these individuals may be from breeding populations located elsewhere in Europe. Some of the Blackbirds may have arrived from Germany, joining Siskins from Belgium and Chaffinches from Sweden. Of course, the individuals from these different populations look virtually identical and we only know of their origins thanks to the efforts of highly skilled volunteer bird ringers participating in the National Ringing Scheme.

Since ringing first began in Britain in 1909, some 32 million birds have been ringed. Of these, subsequent details of some 600,000 have been reported to the BTO, with each report adding new information about the movements made by individual species. The skills necessary to become a bird ringer can only be learned through practice under the close supervision of experienced ringers. For this reason, ringers undertake a period of training of at least one or two years, during which time they are only allowed to ring birds under supervision. As their skills develop so they progress through a series of permits (effectively a licence to ring birds), with independent assessments at each stage. In this way, the National Ringing Scheme maintains the highest standards of bird welfare and data collection.

A Pied Wagtail having its measurements recorded – all part of the ringing process.

RIGHT: BTO bird rings are made from lightweight, durable metal and marked with a unique number and contact address.

RIGHT: A Reed Warbler caught in a mist nest, prior to being ringed.

BELOW: Good, old fashioned birdwatching also provides the BTO with valuable information for its database.

RINGING AS AN ECOLOGICAL TOOL

As well as telling us about the migration and movements of the various bird species visiting Britain and Ireland (*see* pp.24-5), ringing also highlights areas that are important for birds at different times of the year. For example, many migrants make stopovers to rest and replenish energy reserves during each leg of their annual migration; knowing where these sites are is essential if we are to protect birds throughout their life cycle. Ringing can also be used to help us understand changes in the numbers of summer migrants returning here to breed. For example, analyses using ringing data have shown that a great deal of the variation in the numbers of Sedge Warblers breeding here each summer can be explained by the pattern of rainfall in the Sahel region of West Africa, where they spend the winter. In years when there is little or no rainfall, fewer Sedge Warblers return to breed in Britain because many will have succumbed to the effects of drought. Increasingly, ringing data have been used to examine changes in the survival rates of bird species (*see* pp.26-7), highlighting whether it is changes in survival rates or productivity that have driven population declines.

Sedge Warbler – one of the lucky ones to make it back to Britain.

BIRD BIOLOGY

Birds are instantly recognisable for what they are. Many features of their structure and biology influence or determine their appearance and the way they behave.

Like all birds' eggs, this Chaffinch egg has a hard, chalky shell designed to protect the developing embryo inside.

WHAT IS A BIRD?

Birds are one of the most familiar animal groups, their relative uniformity in size and structure making them instantly recognisable. While birds show a tremendous variation in colour, plumage characteristics and bill shape and size, the shared feature of having a body covered in feathers underlines a common ancestry. Another characteristic that binds together most, though not quite all, bird species is the ability to undertake powered flight. Powered flight is an energetically expensive behaviour and, as such, has to be carried out as economically as possible. There are a limited number of ways in which to achieve this economy and this is why most birds show such similarity in their size and structure. Of course, there are some birds that have lost the ability to fly and, with the associated constraints removed, these species often show a very different body plan. For example, Ostrich's primary means of locomotion is on foot and so it has developed powerful leg muscles and attains a greater body weight. One final feature shared by all bird species is that of reproducing by laying eggs (see pp.14-15). This approach to reproduction has a bearing upon the ability to fly, since allowing your embryo to develop inside an egg means that you do not have to carry it around with you for the full period of its development.

Despite variations in appearance, size and habits, all our garden birds share the same covering of feathers and an ability to fly.

Blue Tit

Tawny Owl

Green Woodpecker

ADAPTED FOR FLIGHT

Flight imposes an upper limit on body weight (around 15kg in modern birds) and those species that approach this weight have noticeable difficulty in getting airborne (picture a swan trying to take off). As such, the energetic costs of powered flight favour a body plan where weight is kept to a minimum. The internal skeleton contributes greatly to overall body weight and so most birds have hollow, lightweight bones. Generally speaking, it is the larger bones (limb bones and skull) that have been lightened in this manner but in some of the larger bird species the smallest bones are also hollow. The avian beak is another weight-saving adaptation since it is significantly lighter than the heavy jawbones carried by mammals. Body weight can also be minimised by reducing the size of non-vital organs within the body. For example, the reproductive organs of many bird species shrink in size outside of the breeding season, only attaining their proper size when needed. The bird's musculature also shows a number of significant adaptations for flight. As you might expect, the main flight muscles are large, delivering the power needed to drive the wings. At the same time, the bird is able to maintain a high degree of manoeuvrability because these muscles are situated close to the body cavity. The muscles are responsible for the wing's down-stroke while ligaments drive the upstroke.

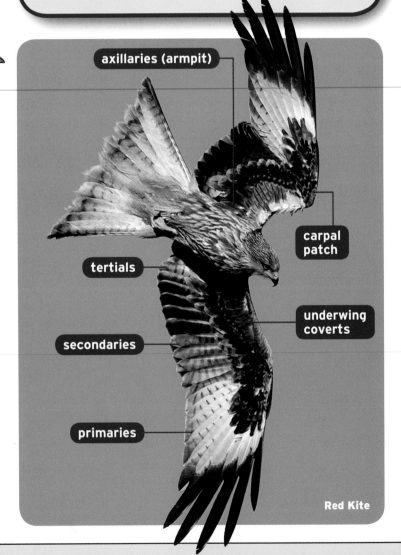

axillaries (armpit)

carpal patch

tertials

underwing coverts

secondaries

primaries

Red Kite

FLIGHT AND FEATHERS

Various tracts of feathers have also been adapted for flight, particularly those along the rear edge of the wing and on the tail. These feathers are elongated (when compared to other body feathers), strengthened and shaped so that, collectively, they can provide the thrust, lift and manoeuvrability needed in the air. The most energetically costly component of flight is the take-off and if you watch small birds you will see that they invariably jump vertically into the air before making a powerful downstroke with their wings to give them the lift needed to get fully airborne. Large birds are just too big to take off in this manner and so have to run along the ground (or water), flapping their wings, until they reach a speed at which they can gain the necessary lift. Once airborne, the process of flying becomes more efficient and many birds can cover a considerable distance with relatively little expenditure of energy. Some species further enhance their energetic effectiveness by soaring on thermals.

Woodpigeon primary (outer flight feather).

Woodpigeon tail feather.

FEATHERS

The plumage of a bird has a number of different functions. As well as giving the bird its external appearance, the plumage can indicate sex and social status or provide camouflage from potential predators. The individual feathers that make up this plumage can even be used to carry water (as seen in sandgrouse) or to make a sound (as happens in Snipe and Woodcock). Feathers are made from keratin, the same substance that forms our nails and hair. Keratin is an amazing material, being light, durable and waterproof. In the form of feathers it is a good insulator, trapping a layer of air between the body feathers and the skin beneath them. Each feather grows from the base, typically growing by 1–2mm per day but reaching up to 13mm per day in some exceptional cases. Once the feather has finished growing it is comprised of entirely dead material and further changes require the growth of a completely new feather (either to counter the effects of wear and tear or to allow for a change in the functional role of the feather). Each bird has a number of different feather types,

Woodpigeon contour feather.

serving different functions within the overall plumage. These include the flight feathers, which are stiff and elongated, held together along the line of the feather shaft by a series of barbules that effectively 'zip' the feather together. Then there are the contour feathers that cover the body, comprising a 'zipped' outer section and an unzipped fluffy inner section; the outer section provides protection, the looser inner section insulates the bird. Other feather types have more specialised functions.

BIRD TOPOGRAPHY - NAMING THE PARTS

One of the best ways in which to improve your bird identification skills is to become familiar with the different parts of a bird. Each of the various feather tracts has its own name and knowing these will help you to understand the descriptions given in field guides and also to take notes that another birdwatcher will be able to understand.

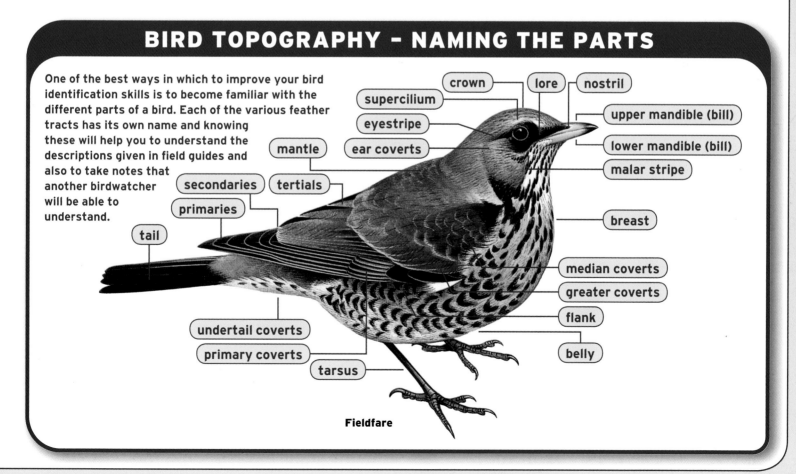

crown · lore · nostril · supercilium · eyestripe · ear coverts · upper mandible (bill) · lower mandible (bill) · malar stripe · mantle · secondaries · tertials · primaries · breast · tail · median coverts · greater coverts · flank · undertail coverts · primary coverts · belly · tarsus

Fieldfare

BREEDING BIOLOGY - EGGS

Most British birds breed in late spring and early summer, which is when you may find an abandoned egg on your lawn or discover an unknown nest in a shrub. Being able to identify the eggs and nests of different species will help you establish which birds are breeding in and around your garden.

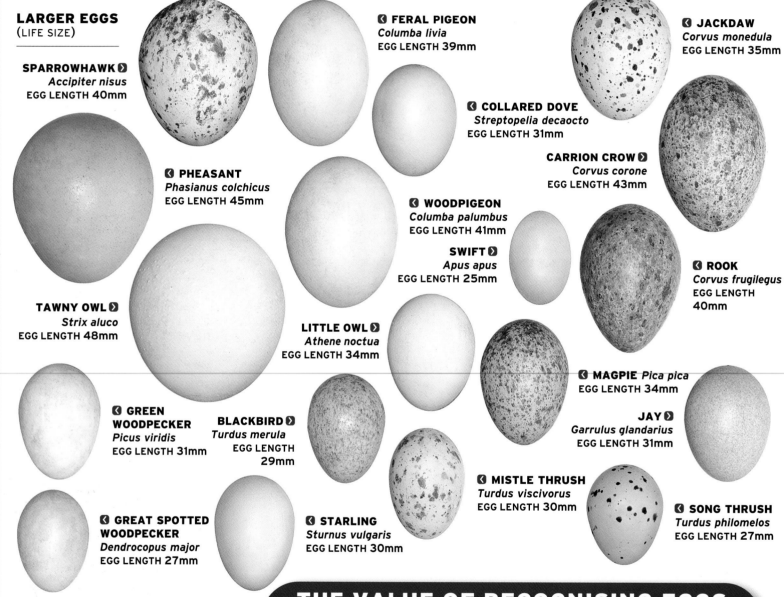

LARGER EGGS
(LIFE SIZE)

SPARROWHAWK
Accipiter nisus
EGG LENGTH 40mm

PHEASANT
Phasianus colchicus
EGG LENGTH 45mm

TAWNY OWL
Strix aluco
EGG LENGTH 48mm

GREEN WOODPECKER
Picus viridis
EGG LENGTH 31mm

GREAT SPOTTED WOODPECKER
Dendrocopus major
EGG LENGTH 27mm

FERAL PIGEON
Columba livia
EGG LENGTH 39mm

COLLARED DOVE
Streptopelia decaocto
EGG LENGTH 31mm

WOODPIGEON
Columba palumbus
EGG LENGTH 41mm

SWIFT
Apus apus
EGG LENGTH 25mm

LITTLE OWL
Athene noctua
EGG LENGTH 34mm

BLACKBIRD
Turdus merula
EGG LENGTH 29mm

STARLING
Sturnus vulgaris
EGG LENGTH 30mm

JACKDAW
Corvus monedula
EGG LENGTH 35mm

CARRION CROW
Corvus corone
EGG LENGTH 43mm

ROOK
Corvus frugilegus
EGG LENGTH 40mm

MAGPIE *Pica pica*
EGG LENGTH 34mm

JAY
Garrulus glandarius
EGG LENGTH 31mm

MISTLE THRUSH
Turdus viscivorus
EGG LENGTH 30mm

SONG THRUSH
Turdus philomelos
EGG LENGTH 27mm

THE VALUE OF RECOGNISING EGGS

By being able to identify the eggs and nests of garden birds you can satisfy your own curiosity about which species are nesting in your garden, but more importantly you can contribute to an important national project, the BTO's Nest Record Scheme. The scheme gathers vital information on the productivity of the UK's birds, enabling researchers to determine if changes in breeding success are behind recorded population declines. Roughly 30,000 nest records, covering some 160 or so species, are submitted to the BTO annually by dedicated nest recorders. Records from gardens are just as important as those from the wider countryside. It is important to minimise nest disturbance for ethical and scientific reasons: all nest recorders operate under a strict code of conduct that ensures the safety of the nest is not jeopardised and this is detailed in the Nest Records Starter Pack available from the Nest Records Unit at the BTO (email: nest.records@bto.org); you can find out more at www.bto.org/survey/nest_records.

BTO Nest Record Scheme

SMALLER EGGS (TWICE LIFE SIZE)

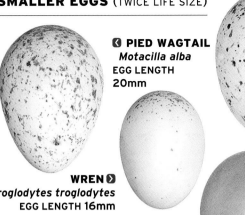

◀ PIED WAGTAIL
Motacilla alba
EGG LENGTH
20mm

WREN ▶
Troglodytes troglodytes
EGG LENGTH 16mm

◀ SWALLOW
Hirundo rustica
EGG LENGTH 20mm

◀ HOUSE MARTIN
Delichon urbicum
EGG LENGTH 19mm

GOLDCREST ▶
Regulus regulus
EGG LENGTH 14mm

**SPOTTED
FLYCATCHER ▶**
Muscicapa striata
EGG LENGTH 19mm

◀ CHIFFCHAFF
Phylloscopus collybita
EGG LENGTH 15mm

DUNNOCK ▶
Prunella modularis
EGG LENGTH 19mm

BLUE TIT ▶
Cyanistes caeruleus
EGG LENGTH 16mm

◀ ROBIN
Erithacus rubecula
EGG LENGTH 20mm

◀ GREAT TIT *Parus major*
EGG LENGTH 16mm

◀ WHITETHROAT
Sylvia communis
EGG LENGTH
18mm

LONG-TAILED TIT ▶
Aegithalos caudatus
EGG LENGTH 14mm

◀ COAL TIT
Periparus ater
EGG LENGTH 15mm

◀ NUTHATCH
Sitta europaea
EGG LENGTH
20mm

MARSH TIT ▶
Poecile palustris
EGG LENGTH 16mm

◀ BLACKCAP
Sylvia atricapilla
EGG LENGTH 20mm

◀ TREECREEPER
Certhia familiaris
EGG LENGTH 16mm

CHAFFINCH ▶
Fringilla coelebs
EGG LENGTH 18mm

HOUSE SPARROW ▶
Passer domesticus
EGG LENGTH 22mm

◀ GREENFINCH
Carduelis chloris
EGG LENGTH 20mm

◀ TREE SPARROW
Passer montanus
EGG LENGTH 20mm

GOLDFINCH ▶
Carduelis carduelis
EGG LENGTH 17mm

◀ SISKIN
Carduelis spinus
EGG LENGTH 16mm

LESSER REDPOLL ▶
Carduelis cabaret
EGG LENGTH 17mm

◀ LINNET
Carduelis cannabina
EGG LENGTH 18mm

◀ BULLFINCH
Pyrrhula pyrrhula
EGG LENGTH 19mm

REED BUNTING ▶
Emberiza schoeniclus
EGG LENGTH 20mm

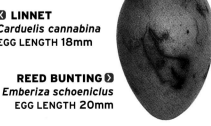

EGG COLLECTING

It is a depressing thought that there are still people who collect the eggs of wild birds, often targeting species that are particularly rare or threatened. Such behaviour is not only damaging to our bird populations but it also undermines important scientific work and conservation efforts. The law is very clear on the issue of egg collecting and anyone in possession of the egg of a wild bird is breaking the law. Egg collectors now face tougher sentences and increased fines but, despite this, there are still a number of hard-core collectors active within the country.

NESTS

Although birds show a great diversity in the structure, placement and construction of their nests, the majority of those built by garden birds are broadly similar in nature. Most garden birds are either cavity nesters (for example, Blue Tit, Starling and Great Spotted Woodpecker) or open nesters (for example, Blackbird, Song Thrush, Chaffinch, Collared Dove). While the nests of cavity nesters are typically less well constructed (because they receive protection from the cavity in which they have been placed), those of open nesters are invariably beautifully woven from small sticks, fine grasses and plant stems. Although it can sometimes be difficult to determine exactly which species has built a particular nest, you should be able to gain a fair idea by looking at the size of the nest, the materials used in construction and how it is lined.

PIGEONS AND DOVES

Woodpigeons and Collared Doves tend to build very simple nests, made from a loosely woven platform of sticks. A Woodpigeon's nest is usually better constructed and with more sticks involved than that of a Collared Dove, whose nest is rather pathetic and often sited on a man-made structure.

THRUSHES

The nests of our larger thrushes (Blackbird, Song Thrush and Mistle Thrush) take the form of a large cup of woven grass and small twigs, with mud used as part of the lining. Those of Blackbird and Mistle Thrush are finished with finer grasses, such that the mud or earth lining is incorporated into the structure and not normally visible within the nesting cup itself. However, the mud lining of a Song Thrush nest is there for all to see, since no finer grasses are added as a final lining to the nest cup.

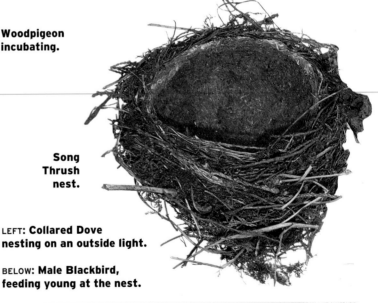

Woodpigeon incubating.

Song Thrush nest.

LEFT: **Collared Dove nesting on an outside light.**

BELOW: **Male Blackbird, feeding young at the nest.**

TITS

While the true tits are cavity nesters, the Long-tailed Tit builds an incredible domed nest, usually positioned in a thick or thorny bush. The Long-tailed Tit nest is made of mosses and lichens and then lined with up to 2,600 feathers collected from the local area. Cavity-nesting Blue Tits and Great Tits make a nest that fills the base of the nesting cavity, usually using a mixture of mosses and grasses (Great Tit tends to use less grass in the construction of its nest than Blue Tit). Like many other cavity nesters the nest cup used by Blue Tits and Great Tits is placed up against the side of the cavity, usually against the wall furthest from the cavity entrance; in compact nestboxes, however, the whole of the box floor is utilised.

FINCHES

As you might expect, the finely woven nests of the various finches are smaller and more delicate than those of the thrushes. While some may be built upon, or incorporate, a base of small twigs (such as those of Bullfinch and Greenfinch), others are woven onto the fork or branch on which they have been placed (for example, Chaffinch and Goldfinch). Perhaps the neatest of these nests is that made by the Goldfinch; compact in form, it is made up of dead grass and rootlets, with smaller amounts of moss, cobwebs and lichen woven in. That of the Chaffinch is similar, although it does tend to include far more moss and have external embellishments in the form of lichen woven with cobwebs.

ABOVE: **Long-tailed Tits at their nest.**
BELOW: **Blue Tit nest in a nestbox.**

Chaffinches at the nest.

Bullfinch pair feeding their chicks.

BIRD BEHAVIOUR

There are many interesting differences between the behaviours exhibited by birds and those of other animals. For example, while birds demonstrate rapid learning ability, they generally do not exhibit the types of 'play' behaviour seen in mammals. Such differences are thought to result from the rapid development of birds. Young birds often attain their adult weight and size in just a few weeks, a duration that may equate to just 1% of their life expectancy. In many mammals, especially primates, adult size and weight are not reached until 30% of the total life expectancy has passed. Hence, the accelerated development rates of birds may favour rapid learning but preclude play. Many different behaviours can be studied by watching garden birds and this approach can reveal a rewarding insight into their lives.

SONG

Song is one of the most obvious bird behaviours, providing a backdrop to our daily lives for much of the year. Song in birds is usually linked to the acquisition and defence of a territory or for mate attraction. Almost all song is generated by the male and appears to be under the control of the male sex hormone, testosterone. In a few species both sexes may sing for all or part of the year (Robin being a good example). Because song is linked to territory and breeding, there is a clear seasonal cycle, the amount of song often declining once breeding is well underway. Many female birds use male song as a means of determining the suitability of a potential mate. In Stonechat, research has found that males with higher rates of singing are better parents, helping to feed the chicks and defend the nest more than males with lower rates of song production.

FLOCKS AND COMMUNAL ROOSTING

While song is a predominantly a feature of summer, flocks and communal roosts are a feature of autumn and winter. Being part of a flock may offer advantages to an individual bird once the restrictions of breeding have been removed. For example, being part of a larger group means that you can spend more time feeding and less time having to watch out for predators, safe in knowledge that other members of the flock will be taking their turn to keep a watchful eye out. Some of the most familiar flocks are those formed by roosting Starlings. These flocks may feature many thousands of individuals, coming together to roost at favoured sites. Small groups of Starlings coalesce into bigger flocks, performing a spectacular aerial ballet before dropping into the roost as darkness falls. Communal roosting may help birds to conserve energy by reducing heat loss. This is why Long-tailed Tits huddle together in a line and Wrens may cram themselves into a nestbox or natural cavity. It is thought that the roosts can also act as information centres, allowing individual birds to assess the condition of their neighbours. The following morning, as the roost breaks up, individuals who have had a poor day feeding can follow those in good condition to feeding grounds where food may be more abundant.

LEFT: **Robin in full song in early spring.**

BELOW: **Starling flock going to roost.**

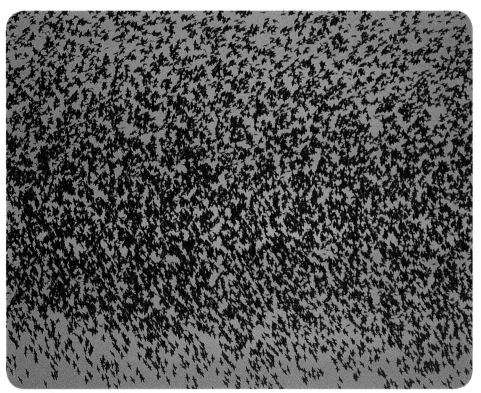

FEATHER MAINTENANCE

It is essential for a bird to keep its feathers in good condition, both for insulation and flight. Routine maintenance involves a number of different behaviours, many of which may be observed in garden birds. Good numbers of birds bathe using water, either splashing themselves in a suitable pool or by dew-bathing against wet vegetation. Typically, a bird will attempt to moisten the surface of the feathers rather than actually soak them. This helps the preen oil to spread across the feathers. Other birds, for example House Sparrow and Pheasant, dust bathe. Preening typically follows on from bathing and, in many species, this involves working preen oil into the feathers. The oil is produced in a preen gland located just above the tail. Preening is also used to manipulate the feathers, realigning the individual filaments that hook together to give each feather its shape. Birds can sometimes be seen 'nibbling' their flight feathers, carefully working each one in turn. Individual birds may occasionally be observed sunbathing, spreading out their wing and tail feathers, and fluffing up their body plumage, to force feather parasites out.

AGGRESSION AND DISPLAY

Most birds use display as a means of settling disputes and maintaining a dominance hierarchy. Neighbouring Robins will advertise ownership of their respective territories through song, escalating this to a series of display postures if necessary. In these, the territory-holding Robin will attempt to perch above the intruder, displaying the red breast to maximum effect. If caught below an intruder, the Robin will throw his head back in an attempt to display his red breast up towards the intruder. Only if these signals fail to repel the intruder will a fight take place. Display is therefore used to settle arguments where possible without the risk of injury that might come with direct confrontation. Displays of aggression may also be witnessed between different species and, if you keep a watch on your garden feeding station, you will soon discover which species are dominant over others.

This Blue Tit failed to read the warning signs and is on the receiving end of a vicious jab from a pugnacious Greenfinch.

ABOVE: **Kingfisher preening**; BELOW: **Robin sunbathing.**

THE DAILY CYCLE

It is soon apparent from watching garden birds that some species arrive to feed early in the morning, while others put in an appearance much later in the day. As research by the BTO has shown, these differences have a biological basis and are linked to physiology, ecology and geographic location. Small birds store less fat for use overnight than larger birds and also have a higher metabolic rate, because of greater heat loss. This means that these individuals are more energetically stressed come dawn and so need to feed sooner in order to make up the shortfall. Larger birds, which also tend to have the competitive advantage at garden feeding stations, can afford to arrive later. Such patterns may also be modified by overnight temperatures, the presence of predators and roosting behaviour.

Winter is a challenging time for all birds, but particularly for small species like this Blue Tit.

House Martins specialise in catching day-flying insects, especially flies; activity of their prey largely dictates the birds' feeding patterns and success.

Robins feed their young mainly on insects and find most of their prey when it moves.

THE DAILY CYCLE IN WINTER

Most species show three peaks to their levels of feeding activity at garden feeding stations during a typical winter day. The first occurs early in the morning, with individuals arriving from their overnight roosts to replace lost energy reserves. A later peak, occurring towards the end of the day, probably has a similar purpose, with individuals topping up their energy reserves before going to roost. The absence of birds in late morning and early afternoon suggests that small birds are balancing the benefits gained from feeding against the increased risks of predation that may result from gathering together to feed. Interestingly, there is a third peak (small in size) that occurs in the middle of the day. While this peak does not fit what might be predicted from the predation-risk model, it may be a consequence of competition between species for access to food. Small or subordinate individuals may be forced to feed at a time that is not ideal because larger or dominant individuals exclude them from feeders at better times of day. A complicating factor in all this is that the food supply may not always be predictable. While the predictable food supply provided by a garden feeding station may afford birds the opportunity to feed at a time of their choosing, where food is unpredictable they may be more likely to feed at the first opportunity possible.

THE DAILY CYCLE IN SPRING AND SUMMER

With broods of hungry mouths to feed during the breeding season, parent songbirds are usually active from when the first glimmer of light appears on the horizon until dusk. Most feed their young, to a greater or lesser extent, on invertebrates and their foraging success, and choice of prey, may vary according to the time of day and the weather. For example, earthworms are generally easier to find around dawn, when the ground is wet with dew, or after rain. By contrast, day-flying insects such as butterflies are only active and easily caught on sunny days between 9am and 5pm.

Although the daily cycle during the breeding season is still under the influence of food availability, it is additionally influenced by breeding biology. Breeding birds have a wider range of needs to balance at this time of the year and the daily routine may vary depending on the stage of the breeding cycle and its associated behaviours. One of the most interesting aspects of the daily cycle is the timing of egg-laying. Most small birds produce one egg per day during the period of egg-laying. In most garden birds the egg is produced around sunrise but in Woodpigeon laying takes place early in the afternoon and in Pheasant it takes place in the evening. Female Blue Tits roost in the nest cavity overnight and usually lay their daily egg at first light just before leaving the nest. Blue Tit incubation starts when a full clutch has been laid.

DAYTIME SONGSTERS

Although the peak of vocal activity occurs first thing in the morning, many bird species continue to sing throughout the day, particularly at the start of the breeding season or if they have failed to find a mate. Whitethroats, for example, sing off-and-on throughout the day throughout May and June.

Singing male Whitethroat.

Heralding daybreak, this male Robin sings to proclaim territorial ownership.

EARLY RISERS

The BTO's Shortest Day Survey looked at the time that birds arrived at garden bird tables on winter mornings. Carried out on the shortest day of the year (hence the name), the survey revealed that Blackbirds were, on average, the first birds to arrive in the morning, appearing some 15 minutes after daybreak. They were followed by Robins (20 minutes after daybreak) and Blue Tits (23 minutes after daybreak). The study also showed that bird species with large eyes (relative to body size) arrived earliest and that individuals of the same species arrived later at urban gardens than rural ones.

THE DAWN CHORUS

The dawn chorus is perhaps the most familiar component of the daily cycle at this time of the year. This chorus of song usually begins before dawn and lasts for some time before ending, often abruptly. Many researchers consider the dawn chorus to represent a trade-off between the conflicting needs to defend a territory and to find food, with fitter birds able to sing for longer. Various explanations have been put forward to explain the timing of the dawn chorus. Research has shown that the chorus occurs at a time when the level of territorial intrusions is at its peak and it is also the time when low light levels may limit feeding opportunities and reduce the effectiveness of visual displays of territorial ownership. Interestingly, bird song can sometimes be heard at night, especially in urban and suburban areas with street lighting. These songsters will usually turn out to be Robins and Blackbirds, species with large eyes and able to cope with the reduced light levels sufficiently well to indulge in a spot of singing. They will not be Nightingales, which are birds of scrubby woodland.

The early bird catches the worm – or at least that's the idea.

ANNUAL PATTERNS

Many garden birdwatchers are well aware of the comings and goings within their own gardens, as different birds make use of the garden habitat and its resources at particular times of the year. However, it is only through the BTO Garden BirdWatch survey that we have been able to quantify such patterns and to determine their ecological basis. For some species the seasonality of garden use is really obvious. Winter visitors, like Redwing and Brambling, stand out because they do not breed here and so are clearly seasonal in their use of gardens. However, many species that may be regarded as 'resident' have a migratory component arriving here in winter. This means that those Blackbirds visiting your garden in December may be drawn from as far afield as central Europe or even Russia.

Redwing **Blackbird**

It's obvious that this Redwing is a migrant visitor but what about the Blackbirds in your garden in winter?

THE AUTUMN TROUGH

The weekly records from the BTO Garden BirdWatch show a distinct autumn trough in the reporting rates of several resident species from gardens. This trough tends to be roughly consistent in its timing and magnitude from one year to the next, matching the period when there is plenty of natural food available within the wider countryside. With an abundance of food available, birds become less reliant on the supplementary food put out in gardens. It is also a time of the year when many garden birds undergo their annual moult, replacing worn-out feathers in readiness for the winter ahead. During the process of moult, most species become shy and retiring, as they attempt to avoid the unwanted attentions of predators while their ability to fly is compromised.

MIGRANTS AND WINTER VISITORS

Summer migrants and winter visitors also show a pronounced seasonality to their use of gardens, which matches the period when they are either in Britain breeding, or passing through on migration. The pattern shown by summer migrants is likely to be more consistent from year to year than that of winter visitors, since the timing of arrival of many winter visitors is linked to food availability and weather conditions elsewhere. One of the best examples of the magnitude of variation in the peak reporting rate between years can be seen in the Brambling. This species winters in Britain and Ireland in varying numbers, depending largely upon the availability of Beech mast. In years with a poor crop of mast the BTO Garden BirdWatch reporting rate is higher but in years with a good crop it is much reduced. The timing of arrival of winter visitors can also be dependent upon weather conditions. Cold-weather influxes of Starlings, both new arrivals from continental Europe and birds moving into gardens from the wider countryside, shows as mid-winter peaks to the BTO Garden BirdWatch reporting rate.

Migrating Brambling flock.

RESIDENT BIRDS

Other species resident in Britain and Ireland and associated typically with the wider countryside may also show a seasonal pattern to their use of gardens. Species like Yellowhammer and Reed Bunting, which are birds of farmland, are reported from gardens only rarely throughout much of the year. However, the reporting rate increases briefly during late winter, with birds moving in to use feeding stations placed in rural gardens. The timing of this peak suggests that these seed-eating birds are turning to gardens at a time of the year when natural seed supplies are exhausted. A number of other species show increased use of gardens at particular times of the year. The reporting rate for Jackdaw peaks in early summer, a period when the adults find it particularly difficult to find sufficient food for their chicks. Great Spotted Woodpeckers show a similar peak, this time with adults bringing their chicks to garden feeding stations to make use of hanging peanut feeders, a useful food for those chicks still learning the ropes.

Yellowhammer - seen in the garden mainly in late winter.

DIFFERENT GARDENS, DIFFERENT PEAKS

While the BTO Garden BirdWatch reporting rates for many species are similar in their pattern across different types of garden (rural, suburban, urban), there are some interesting differences. Perhaps the most interesting is that for Blackcap. Although Blackcaps are predominantly summer visitors to Britain and Ireland, they have started to winter here in increasing numbers (*see* pp.158-9). Urban and suburban gardens show higher winter reporting rates than rural gardens, a pattern that reflects the availability of suitable food at garden feeding stations and the fact that urbanised areas are, on average, several degrees warmer than the surrounding countryside.

Female Blackcap, feeding on blackberries. Once natural food sources have been used up in the garden, feeding stations become more attractive to the species.

SEASONAL PATTERNS IN THE USE OF GARDENS

The following graphs show how the use of gardens varies throughout the year.

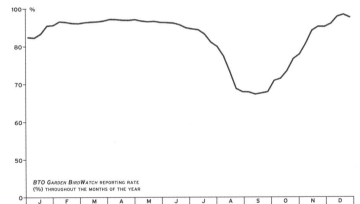

BTO *Garden BirdWatch* REPORTING RATE (%) THROUGHOUT THE MONTHS OF THE YEAR

Blackbird – there is a pronounced autumn trough in garden use.

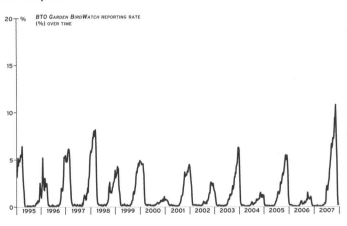

BTO *Garden BirdWatch* REPORTING RATE (%) OVER TIME

Brambling – this winter visitor arrives in gardens in greater numbers in some years than others.

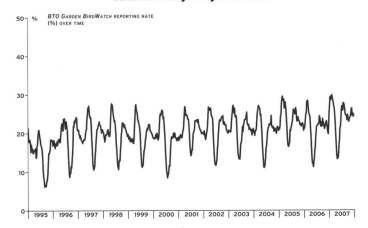

BTO *Garden BirdWatch* REPORTING RATE (%) OVER TIME

Great Spotted Woodpecker – they make greatest use of gardens in the summer, often bringing their youngsters with them.

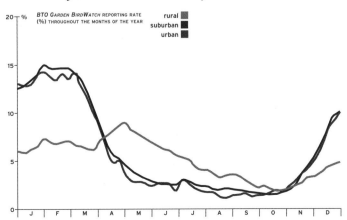

BTO *Garden BirdWatch* REPORTING RATE (%) THROUGHOUT THE MONTHS OF THE YEAR

rural
suburban
urban

Blackcap – they make greatest use of urban and suburban gardens during the winter but rural gardens in summer.

MOVEMENTS AND MIGRATION

Birds have a remarkably ability to cover long distances and to cross areas of unsuitable habitat in search of more favourable conditions elsewhere. The ability to fly frees them from many of the constraints faced by other groups of animals that may limit their populations and movements. The movements made by birds fall into a number of categories.

MIGRATION

Migration is the most extreme form of movement made by birds, taking them on a predictable and (largely) annual flight between two distinct areas, one used for breeding and the other used for wintering. Such migrational movements result from the fact that most avian food resources, notably insects, are strongly seasonal in their abundance. Most of our summer migrants are insectivorous and breed in Britain and Ireland during the long daylight hours that our northern summers provide. Insect abundance declines with the approach of winter and the insect eaters are forced to retreat south to latitudes where insects remain readily available. Other migrants visiting Britain and Ireland arrive for the winter, notably those avoiding colder weather further north and east. Migration is seldom an all or nothing process and some populations or species undergo a partial migration, where some individuals migrate while others remain here. Typically, partial migrants are responding to food availability, competition for that food and local weather conditions.

HOW WE KNOW WHAT WE KNOW

Much of our knowledge of how and why birds move comes from the work carried out by bird ringers, but the occurrence of such movements can also be seen in the data collected by projects like the BTO Garden BirdWatch and the BTO-led BirdTrack project. With new technologies now available, researchers have also been able to track the movements of individual birds through the use of radio-transmitters and satellite tags, the latter providing an amazing insight to the routes taken by some of our larger migratory birds, like Osprey and Whooper Swan.

CHAFFINCH

In winter, our resident Chaffinches are joined by immigrants from Scandinavia and eastern Europe.

SWALLOW

Thanks to the efforts of BTO ringers we have a good understanding of Swallow migration. Our Swallows migrate south through western Europe, cross the Sahara and winter in South Africa.

The Cuckoo is one of the few species to migrate southeast from Britain in the autumn, as revealed by ringing studies. Their migration continues south of the area covered by this map, to wintering grounds in Africa.

CUCKOO

IRRUPTIONS AND ERUPTIONS

For some species the availability of favoured foods may not simply change with season but may also vary (often dramatically) from one year to the next. This is particularly true for those species that feed upon berries (e.g. Fieldfare and Waxwing), tree seeds (e.g. Chaffinch) or small mammals (e.g. owls and raptors). Under certain conditions, trees or shrubs may, over very large areas, produce very few fruits or seeds in a particular year. This forces those birds that would normally feed on these fruits and seeds to move into other areas in search of food. These movements, known as eruptions (where the birds leave an area) and irruptions (where the birds arrive at an area), can result in the periodic arrival in gardens of Waxwings, Fieldfares, Siskins and Bramblings, sometimes in very large numbers.

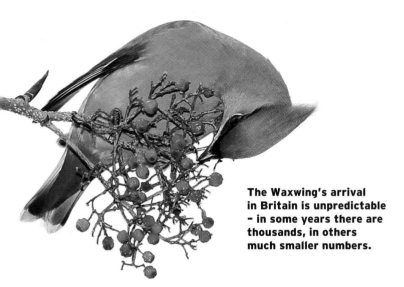

The Waxwing's arrival in Britain is unpredictable – in some years there are thousands, in others much smaller numbers.

NAVIGATION

Migration is a complex process, not least because each migrating bird will need to navigate from its breeding ground to where it is going to spend the winter. The bird has to know where it is going and the route it will take to get there.
In young birds, this journey will take it to a place it has never been before. Birds have been shown to use a range of different mechanisms to help them navigate: some use a star compass, some a sun compass and others use the Earth's magnetic field or a combination of methods. As well as a requirement to follow a route, the individual bird has to allow for the effect of weather conditions and be able to reorientate itself should it be blown off course.

Strong easterly winds forced this spring migrating Reed Warbler off course – it ended up on a seacliff in Devon.

DISPERSAL

Even among birds that we think of as sedentary residents, during the first few months after fledging, young birds usually move away from where they were born, a process known as 'dispersal'. By moving away from where they were born the young birds are lessening the chances of competing with their parents for limited resources like nest sites or food. In addition, such movements reduce the degree of inbreeding, something that is further accomplished by the fact that young females typically disperse a greater distance than young males. Although these dispersive movements may be many kilometres, in most species they are much shorter. For example, the average distance that a Great Tit will move between where it was born and where it will breed is just six or seven territory widths – not much at all.

This young Whitethroat ended up on the Isles of Scilly in October, when really it should have been in Africa. Who knows whether it ever completed its journey?

LIFE AND DEATH

Garden birds face many hazards during the course of their short lives, so it is not surprising to discover that their mortality rates are high. Individuals have to find sufficient food, face predators and contend with a range of man-made hazards.

LONGEVITY AND SURVIVAL RATES

One of the most commonly asked questions is 'how long do birds live?' This is not a straightforward question to answer because scientists studying birds tend to think in terms of populations rather than individuals. As such, they use the term survival (relating to the chances of an average bird surviving from one time period to another) rather than longevity. Although there is published information from bird ringing on the maximum recorded difference between the time at which a bird was fitted with a ring and when it was found dead, the information from bird ringing is usually used to calculate annual survival rates. These can best be explained by considering a population of 100 Blue Tits, all born in the same year. It might be that just 20 of these individuals survive their first winter, giving an annual survival rate of 20%. Ten of these 20 birds might then survive through to the next year, giving an annual survival rate for the second year of 50%. Note that the survival rate has increased – because young, inexperienced birds tend to suffer greater levels of mortality than more experienced birds. The annual survival rate for the third year may remain at 50%, leaving only five of our original 100 birds alive. If this rate continues, then the number of survivors will drop to two or three individuals the following year, with the survivors also perishing within another couple of years. Survival rates can be calculated separately for males and females or for different phases of a bird's life. This allows scientists to understand the critical stages of a bird's life and to identify mortality factors that may be important for a particular age, sex or life stage. Researchers at the British Trust for Ornithology use models that take these survival rates and use them to find out why particular species are in decline.

AGENTS OF MORTALITY

Although birds may gain a number of advantages from living within gardens, there are particular hazards associated with gardens that may be more significant than in other habitats. It is possible to get some idea of the relative importance of these different hazards from bird ringing, by looking at the reported circumstances under which birds have been found dead within gardens. However, these need to be treated with care, since some finding circumstances (e.g. being killed by a cat or by flying into a window) may be more obvious than others (e.g. dying from old age or disease). The two most important man-made hazards facing garden birds appear to be motor-traffic and windows but there are others, such as coming into contact with noxious chemicals or becoming entangled in netting.

ABOVE: **These Great Tits have chosen to nest in a garden nestbox – what are their chicks' chances of survival? Thanks to its research, the BTO has the answer.**

LEFT: **BTO research and statistical analysis allows the survival probability of this juvenile Blue Tit to be estimated with a fair degree of precision.**

IN THE FIRING LINE

It appears that the most significant predator of garden birds is the domestic cat and a number of studies have attempted to quantify the impacts of cat predation on bird populations. While one study found that at least one-third of all House Sparrow deaths in a Bedfordshire village were due to cat predation, it should be stressed that we do not know what effect this level of mortality had on the House Sparrow population itself. If food availability were the main factor controlling overwinter survival and population size in the House Sparrow population, then the high level of cat predation may simply have reduced competition for food and enabled more House Sparrows to survive the winter. Under such circumstances the level of cat predation witnessed may not have had any impact on the size of the House Sparrow population. More work is clearly needed in this area. Sparrowhawks also take garden birds and appear where small birds are abundant. Unlike the domestic cat, the Sparrowhawk is a native predator and part of the natural system, its population effectively controlled by the availability of food. Despite detailed studies having been carried out to investigate the link between the recovery of Sparrowhawk populations and the decline of several songbird species, there is, as yet, no scientific evidence to support the argument (put forward by some) that Sparrowhawks have caused songbird populations to decline. There are several species that may predate the nests of garden birds, including Magpie, Jay, Great Spotted Woodpecker and Grey Squirrel. While each of these predators may wreck individual nesting attempts, their impact at the national level is not thought to be significant.

ABOVE: Magpies are undeniable nest-robbers but there is no scientific evidence to suggest they have a significant impact on songbird numbers. And, being native, they too are a part of our avian heritage.

RIGHT: Life for a songbird is a question of survival of the fittest. Sparrowhawks, like other predators, invariably target weak or inexperienced prey – this is evolution in action.

BELOW: Caught in the middle, House Sparrows, like all small birds, are under attack from all sides; natural predators include Sparrowhawks while man-made hazards include domesticated cats and, of course, man's at times devastating impact on the species' environment.

Sparrowhawk

Domestic Cat

House Sparrow – caught in the predatory crosshairs.

THE IMPORTANCE OF GARDENS FOR BIRDS

Within Britain, private domestic gardens occupy some 10% of the available land area, collectively making them an important habitat at the national level. They make a significant contribution to the amount of urban green space and remain a major contributor to urban biodiversity. However, some people still dismiss gardens (within a wildlife context) as 'artificial', full of introduced plants that are maintained by intrusive management practices and far removed from the structural complexity of so-called 'natural' habitats. While it is true that gardens are managed for our own ends, sometimes intensively so, these critics need to take a hard look around them at the British countryside. Most of our 'natural' habitats are also managed – think of a hay meadow or coppiced woodland. Indeed, there are few official nature reserves that are not managed to a greater or lesser degree and in this respect gardens are part of a wider spectrum of land management. It is becoming increasingly clear that gardens really are important for wildlife.

GOOD FOR BIRDS

Birds are one of the most visible components of the garden flora and fauna and perhaps also the group most favoured by Man. Many householders deliberately attract wild birds into their gardens by providing food and, to a lesser extent, nesting and roosting opportunities. Research carried out by BTO volunteers and staff, has shown clearly that gardens, and the food we provide within them, can have a very positive influence on wild bird populations. For example, it has demonstrated that farmland birds like Reed Bunting and Yellowhammer markedly increase their use of garden food resources when their farmland populations are in difficulty. Perhaps garden bird feeding buffers bird populations against the declines taking place within the wider countryside? Whether or not this is the case, there are other ways in which the use of gardens by birds is of great conservation value. Birds seem to be particularly able to make use of garden resources because they are highly mobile and able to cross areas of unsuitable habitat in order to find food, nest sites or shelter.

Robins are an almost guaranteed feature of gardens throughout Britain and Ireland, be they small or large, established or new.

GARDENS AS A HABITAT FOR BIRDS

There are virtually no bird species within Britain and Ireland that occur only within gardens, to the exclusion of all other habitat types, although one might argue that nesting Swifts and House Martins are limited to human habitats. However, gardens have been shown to support a large component of the populations of a number of breeding species, for example Blackbird (33% of our Blackbird population breeds in urbanised habitats), Starling (54% breeding in urbanised habitats) and House Sparrow (62% breeding in urbanised habitats). The BTO's Garden Nesting Survey, established by Richard Bland and John Tully and carried out by BTO Garden BirdWatchers, has emphasised that the size of breeding bird populations within gardens may be substantially greater than previously thought. For some species, such as Blackbird, populations exist at higher densities in gardens than in farmland and are more productive. By contrast, for other species, garden populations may be less productive than their farmland or woodland counterparts (for example Blue and Great Tits, which need vast numbers of caterpillars for their chicks).

Female Blackbird incubating.

NOT ALL GARDENS ARE EQUAL

The bird community of a city centre garden will be very different from that of a large rural property or a suburban 'semi'. Such differences occur for a number of reasons. Some are due to the nature of the garden, while others are the result of factors operating at a wider spatial scale (e.g. geographic location or nearby habitats). An analysis of the BTO Garden BirdWatch dataset has shown that the likelihood of many species occurring in gardens is more dependent upon the nature of the surrounding habitat rather than on features within the garden itself. A rural garden surrounded by arable farmland will be much more likely to feature birds such as Yellowhammer and Reed Bunting in winter than a garden in countryside where grassland dominates. The presence of nearby woodland may increase the number of species using a garden at certain times of year but birds may leave to feed in the woodland when suitable fruits and seeds are available; this results in a noticeable autumn 'trough' in the Garden BirdWatch reporting rate. Gardens situated near the coast (especially the south and east coasts) are the ones most likely to turn up a rare migrant, seeking shelter after completing a long sea crossing.

THE FUTURE OF GARDENS

With all the changes that have occurred in the wider countryside over the past 60 or so years, it is likely that garden bird populations will have changed, both in terms of numbers and in terms of the range of species using gardens. Long-term monitoring projects, like the BTO's Garden BirdWatch and its Garden Bird Feeding Survey, are a means by which the changing importance of gardens can be studied. The introduction of new foods and feeding techniques has enabled some species to increase their use of gardens, while long-term declines in other species may result from problems in other habitats. By monitoring garden birds, their productivity and their populations we should be able to increase our understanding of the importance of gardens and be better placed to implement effective conservation strategies. With increasing demands for new housing (placing pressures on green-field sites and urban green space) it is becoming even more important to find out just how important gardens are for birds.

LEFT: **The increasing popularity of niger seeds as a source of food for birds has encouraged more Lesser Redpolls to visit garden feeders. Who knows what future surprises innovations in feeding will bring for garden birdwatchers.**

BELOW: **It takes no time at all for a modern garden to be transformed into a green oasis, with three-dimensional wildlife appeal.**

DO GARDENS PROVIDE A GOOD BREEDING OPPORTUNITY?

Great Tit breeding success in gardens has been shown to be lower than that in woodland, a consequence of the greatly reduced availability of insect food in most gardens. Yet winter survival rates are higher in gardens because of the supplementary food we make available. The presence of this food actually attracts Great Tits into gardens from nearby woodland during late autumn and winter, especially in those years when woodland seed supplies are low. So Great Tits using gardens in winter may increase their chances of survival over those that remain within woodland. However, those Great Tits that do remain in woodland, and which manage to survive the winter, are more likely to claim the available territories than those that have wintered elsewhere. This means that some of those Great Tits wintering in gardens will be unable to secure a breeding territory in the preferred woodland breeding habitat and may be forced to breed in the less suitable garden environment. This highlights how complex natural systems are, and how the importance of gardens for bird populations is not necessarily something that can be easily determined.

ATTRACTING GARDEN BIRDS

Different gardens attract different birds. Some of these differences are down to where your garden is located in the country, others are due to the nature of surrounding habitats. While an open garden in the Brecklands of west Norfolk might attract a visiting Green Woodpecker, a similar garden in North Wales is more likely to have a visiting Raven. Similarly, while a suburban garden near an old railway line might be visited by a young Whitethroat, a nearby garden in the middle of a housing estate is unlikely to be so fortunate. Although moving the position of your garden is not usually possible, it is easy to provide shelter, food and nesting opportunities to improve your chances of attracting a wider range of species.

Unsurprisingly, if you live within sight of woodland then you are more likely to have Nuthatches visiting your garden than if you live in the middle of a city.

THE AVAILABLE 'SPECIES POOL'

The number of bird species that make up a particular bird community, known as 'species richness', shows significant variation across Britain and Ireland. The greatest levels of species richness during the breeding season occur in the southeast and decline as you move further north and west. Species richness also varies seasonally, with the arrival of migrants from other regions. These summer and winter visitors show different geographical patterns to their settlement, again influencing the local species pool. Many of our summer migrants breed in the southern half of Britain and Ireland and only make it further north and west in smaller numbers, or not at all. The position of the garden in relation to surrounding habitats is particularly relevant in the case of those species that only make use of gardens at certain times of the year, notably during late winter when feeding opportunities elsewhere are limited. Rural gardens tend to be the ones used by farmland species, while those gardens near to woodland are more likely to have visiting Jays, Treecreepers and Great Spotted Woodpeckers.

LOCATION, LOCATION, LOCATION

Observations collected by the BTO's army of Garden BirdWatchers have revealed that Robins are most likely to visit rural gardens, favouring these over suburban and urban gardens. However, the reverse is true for House Sparrow, which shows a clear preference for urban gardens. These differences relate to the requirements of the birds, their ecology and what is available for them in a particular type of habitat.

rural ■
suburban ■
urban ■

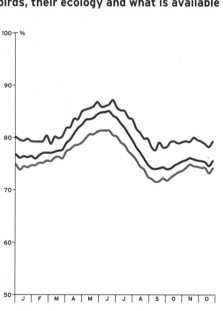

BTO GARDEN BIRDWATCH REPORTING RATE (%) THROUGHOUT THE MONTHS OF THE YEAR
House Sparrow

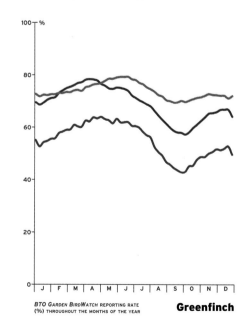

BTO GARDEN BIRDWATCH REPORTING RATE (%) THROUGHOUT THE MONTHS OF THE YEAR
Greenfinch

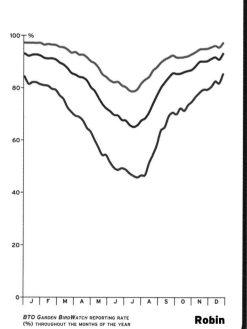

BTO GARDEN BIRDWATCH REPORTING RATE (%) THROUGHOUT THE MONTHS OF THE YEAR
Robin

FACTORS OPERATING WITHIN THE GARDEN ITSELF

Most of the bird species visiting a garden will be looking for habitat features and resources that are similar to those present in other 'natural' habitats. These might include cover from predators, perches from which they can sing and suitable foods. Many gardens will have some of these features but only the largest gardens have the space to hold the diversity of habitat structure favoured by the greatest number of species. Most of our gardens do not have enough of these features and resources to support any given bird throughout the year. This is why many species and individuals are seasonal visitors, arriving at those times of year when resources elsewhere are limited. You can, however, design your garden so that there are more opportunities for visiting birds, and extend the period over which they visit. Other sections in this book show you how to do this.

THE IMPORTANCE OF FOOD AND WATER

The feeding ecology of different birds influences which species will be able to utilise the garden environment. The lack of dense cover, the limited range of plant species present and the use of pesticides often mean that garden habitats support relatively low numbers of insects, spiders and other invertebrates. In turn, this reduces the suitability of the garden environment for insectivorous birds like warblers. The bird species most commonly reported from gardens are ones that either tend to feed on seeds (such as finches) or are omnivorous (such as the Blackbird). A BTO study, looking at factors that influence the way birds use gardens, revealed that the provision of food significantly increased the occurrence of 24 of our most familiar garden species. Water is another important component of a bird-friendly garden. Birds require water for drinking and bathing throughout the year.

RIGHT: **A garden pond will encourage birds from the neighbouring countryside, such as this Whitethroat, to pay regular visits to drink and bathe.**

LEFT: **Grow apples trees in your garden, or scatter windfalls gathered elsewhere, and you may be visited by Fieldfares when times are hard in the countryside at large.**

RIGHT: **It is during the summer months, when other water sources may be scarce, that a birdbath really comes into its own.**

BELOW: **A strict pecking order can be seen among Greenfinches at a feeder.**

FOOD AND FEEDING

The modern approach to garden bird feeding is to use a range of foods that meets the specific nutritional requirements of a wide range of species over the course of the whole year. The traditional approach, of just providing bread, fats, peanuts and a few kitchen scraps, and largely limited to the winter, is increasingly losing favour, as garden birdwatchers make better use of new foods. Most small birds need to consume as much as 30% of their bodyweight in food every day, just to maintain themselves. Feeding modern, high quality and oil-rich foods offers the best opportunity to provide birds with the energy and nutrients they require.

BIRD FEEDERS

Hanging bird feeders come in two main forms, one for seed and the other for peanuts. The choice of an appropriate design can be geared to local conditions. Some cage-feeders are designed to only provide access to smaller birds and these are an obvious choice where Feral Pigeons, Woodpigeons or Grey Squirrels are a nuisance. Similar cages are available for use on bird tables or at ground feeding stations. The diversity of feeders and tables offers the garden birdwatcher plenty of opportunity to develop a garden feeding station that benefits a wide range of species, feeding in many different ways.

LEFT AND BELOW LEFT: **Two alternative types of peanut feeder.**

Both at the bottom of the pecking order, Siskins and Blue Tits will sometimes feed alongside one another on a feeder with a minimum of conflict.

DOES FEEDING MAKE A DIFFERENCE?

There is a wealth of scientific evidence to demonstrate that the provision of supplementary food can make a real difference to birds. Supplementary feeding has been shown to increase both winter survival and productivity in a range of species. Although much of this work has been on bird populations in habitats other than gardens, the same principles apply. Most of the bird species that use gardens are limited by food availability – something that is particularly apparent in winter. The provision of supplementary foodstuffs that reflect the species' 'natural' food, allows the garden birdwatcher to positively influence the productivity and survival of a range of bird species.

Bramblings are occasional and highly prized visitors to the winter bird table. Their 'natural' diet of wayside and woodland seeds is mimicked by the provision of sunflower seeds.

Niger seed feeders have narrow delivery slits to prevent undue spillage of these tiny seeds. Only seedeaters with slender bills, such as Goldfinch (ABOVE) and Redpoll, can take advantage of this food source.

WHEN AND WHAT NOT TO FEED

Both the BTO and RSPB advocate the provision of appropriate foods throughout the year, not just in winter. During late spring and in the breeding season, many species may find it difficult to locate enough food to feed both themselves and their chicks. In many species the growing chicks are fed on invertebrates, the parents only taking supplementary food for themselves. Although adult birds appear able to differentiate between different types of food, and to select appropriate food for their young, some studies have shown that they will also deliver food provided at local bird tables to the nest. The evidence suggests, in Great Tits and Blue Tits at least, that the parent birds may feed their chicks on the supplementary foods when favoured chick foods are in short supply. This means that consideration needs to be given to which foods are provided. For example, whole loose peanuts (which could choke young chicks) should be avoided and we recommend that peanuts are only ever provided behind a wire mesh. Other foods that should be avoided are those with a high salt content (salted peanuts and bacon rind), those that may swell up inside the bird (desiccated coconut), those that are mouldy or spoiled and those that may harbour bacteria (such as meat scraps).

Supplementary food can be provided in many ways, depending upon the type of food being used and the species for which it is intended. Different species are adapted to feeding on a range of foods and have clear preferences in how they obtain their food. Some species are primarily ground feeders (Robin, Dunnock and Blackbird, for example) while others prefer to feed from hanging feeders (Great Tit, Blue Tit, Great Spotted Woodpecker and Siskin).

BELOW: **Two alternative designs for a bird table that incorporate wire mesh peanut feeders into the framework.**

RIGHT: **Drilled holes in this suspended log can be filled with fat and impregnated with mixed seed and grain, satisfying a range of different feeders.**

WHERE TO FEED

It is sensible to place hanging feeders and raised bird tables close to high cover, so that feeding birds can shelter in the bushes if a Sparrowhawk appears. Conversely, the best place for a ground feeding station is away from low vegetation, reducing the danger from cats, waiting to pounce. Birds balance the benefits of feeding on a particular feeder against the risks of attack from a predator. Dominant adult Great Tits have been shown to preferentially feed on feeders close to cover, pushing less dominant individuals onto more exposed feeders, where the risk of Sparrowhawk predation is greater.

Dunnocks will quite happy forage beneath a bird table or feeder, picking up fallen scraps.

FOOD AND FEEDING

Seeds tend to attract a wider range of species than peanuts on their own, and many garden birdwatchers provide black sunflower seeds or sunflower hearts as their staple foods. Alongside these, quality peanut, niger seed and high-energy seed mixes are equally valuable; if you use peanuts it is important that they are endorsed by a wildlife body such as the BTO, BirdCare Standards Association or RSPB and hence have been tested against the presence of a naturally occurring poison called aflatoxin. Sultanas are good for ground-feeding Blackbirds and Song Thrushes, though note these may prove toxic to pet dogs. Finely grated cheese and windfall apples can be very useful, particularly in winter, while peanut cake – a mix of vegetable fats and peanut flour – is good for Long-tailed Tits and Nuthatches.

PEANUTS

Peanuts are high in the oils and proteins needed by birds and have been used for many years at garden feeding stations. Always buy good quality peanuts from a reputable source. Peanuts are best supplied behind a mesh, so that a bird cannot take a whole peanut away, but can also be supplied as granules or peanut cake.

BLACK SUNFLOWER SEEDS

First introduced in the early 1990s, this seed revolutionised bird feeding, by providing a high-energy food in a readily accessible form. These seeds feature in many table seed mixes but can also be presented on their own in hanging seed feeders. They are a favourite of Greenfinches and tits, although they may be shunned if sunflower hearts are available.

SUNFLOWER HEARTS

Although more expensive than the black sunflower seeds from which these hearts are extracted, they have two advantages. First, the birds can feed more quickly because they do not have to remove the husk. Second, the lack of a husk means that there are no unsightly piles of husks that typically appear under hanging seed feeders containing black sunflower seed.

NIGER SEED

A relatively new introduction to the bird feeding market and one that has found favour with Goldfinches, which seem to like the small size of these seeds. Because they are so small, niger seeds need to be supplied in a specially adapted feeder. They are rich in oil and ideal for birds with bills delicate enough to deal with them. They can be mixed with other seed mixes or sprinkled on the ground for Dunnocks.

SEED AND GRAIN MIXES

There is a vast range of seed mixes available on the market and some are much better than others. Cheaper mixes often contain a high proportion of cereal and tend to attract pigeons. Better quality mixes are lower in cereal content and so are particularly suitable for finches, tits and buntings. The best quality mixes are carefully balanced to cater for a range of bird species and their differing nutritional requirements.

FRESH COCONUT

Unlike desiccated coconut, which should not be fed to birds (it causes dehydration), fresh coconut halves are a good source of food for birds, and Long-tailed Tits in particular seem to relish it.

HYGIENE PRECAUTIONS !

Wild birds are susceptible to a range of different diseases, including Salmonellosis and Trichomonosis, both of which can result in the appearance of fluffed-up and lethargic birds at garden feeding stations. Birds can gather in large numbers at garden feeding stations, just as they do at naturally abundant food sources, and this may increase the chances of disease transmission. There are a number of sensible precautions that can be adopted to reduce the risk of disease transmission, thus ensuring that the benefits of garden feeding are maximised:

■ Keep clean the surfaces on which birds feed. Ideally, brush bird tables daily to clear away droppings.

■ If you feed on the ground, do not put food in the same place every day; moving the feeding area around regularly helps reduce the risk of disease transmission.

■ Provide food at several sites in the garden to avoid a concentration of birds in the same place.

■ Move hanging feeders periodically and keep the area beneath them clear of droppings, spilt food and seed husks.

■ Clean bird tables and hanging feeders regularly, ideally using a recognised cleaning agent designed for the purpose. Rinse feeders and tables after cleaning and ensure they are dry before refilling with food.

■ Ensure that any water you provide is fresh, and that birdbaths are cleaned and rinsed regularly.

■ Some bird diseases can be passed on to humans, so it essential that you should observe scrupulous personal hygiene.

WATCHING GARDEN BIRDS

Watching garden birds can be an incredibly rewarding experience and it is one of the main reasons why many people provide food for birds on a daily basis. To watch wildlife from the comfort of your home provides a great opportunity to see species that are significantly more wary when encountered in other habitats. It also allows greater insight into their behaviour and ecology. The degree of interest that garden birdwatchers show varies considerably: many are content to gain pleasure simply by watching the antics of feeding and nesting birds; for others, their interest centres on knowing what the birds are doing and why they are doing it. This is where projects like the BTO Garden BirdWatch can provide an extra stimulus, by encouraging observers to pay greater attention to what is going on in their gardens. Once this interest has been stimulated it can become a serious and time-consuming hobby!

ABOVE: **Passing on the wisdom of age to a new generation, a budding ornithologist is learning the ropes for BTO recording.**

BELOW: **Experienced birdwatchers can distinguish Song Thrush from Redwing by the colour of their underwings.**

Song Thrush

Redwing

GETTING STARTED

One of the great attractions of watching garden birds is that it does not require lots of different equipment, nor does it necessitate a large amount of commitment or expense. The activities of birds using a bird table can be watched over breakfast, while washing-up or from the comfort of an armchair by the fire. Other than the various feeders, bird tables and nestboxes that can be used to attract the birds in the first place, all that is really needed is a pair of binoculars (and even these may not be essential) and a good field guide to help with identification. There are a number of excellent field guides on the market (see Further Reading). Use a camera to record unusual garden visitors or behaviour - this will give you more time to reflect on the question of identification. In addition, photographs can then be shown to county bird recorders, enabling the record to be accepted for county bird reports far more readily than would otherwise have been possible. Some garden birds are quite secretive and may only reveal their presence through their song or calls. It is often more difficult to identify a bird based on its call or song than on sight but there are some very good sound recordings now available. There are also specialist videos and DVDs to help you with the identification of garden birds.

The call of the Green Woodpecker is distinctive enough to allow identification without actually seeing the bird.

TAKING NOTE

One of the delights of watching garden birds is that there is always something going on, whether it is a Coal Tit taking sunflower hearts away from a feeder to store elsewhere, an aggressive encounter between two Greenfinches or a young Blue Tit begging for food from one of its parents. It is worth keeping a diary and daily observations mean that seasonal trends become obvious. In spring, male Robins and Blackbirds will be setting up breeding territories, leading to aggressive encounters between territory holders and intruders. Robins can be particularly intolerant of interlopers, attacking other Robins and even Dunnocks. If you are fortunate enough to have breeding Dunnocks then it is worth looking out for examples of their complex breeding system and amazing pre-copulation behaviour.

The Dunnock's rather undistinguished plumage belies its complex social life and is well worth studying.

SPOTTING THE UNUSUAL

During the autumn, many gardens go quiet as birds feed elsewhere or undergo moult. Moulting birds tend to skulk about but may be seen preening their new plumage. Autumn is also a time of year when migrant birds are on the move and this is when rarities begin to turn up in gardens. Migrant warblers may appear, to feed on fruits, and some fortunate garden birdwatchers will be visited by a Wryneck – the rarest of our woodpeckers. With the arrival of winter, most garden feeding stations become very busy places as the birds seek out the food that they need to see them through the long winter nights. At this time of the year it can prove very interesting to watch the daily pattern of arrivals and departures, the competition between individuals for access to food and the different feeding styles. It is also during the winter months that small birds can be seen roosting together in bushes or in roosting pouches and nestboxes. It is quite a sight to see a succession of Wrens enter a nestbox late on a winter afternoon. Unusual behaviour may be seen at any time of the year and that is part of the enjoyment of watching garden birds, not knowing what you may see next.

RIGHT: A migrant Wryneck is a prize autumn find for any garden birdwatcher.

WHAT TO LOOK FOR IN A PAIR OF BINOCULARS

Binoculars come in a wide range of shapes and sizes, not to mention prices. Between the two extremes - very cheap but not very good, and more expensive than you actually need - there are many models that will provide you with a lifetime of good service. When choosing a pair of binoculars, ensure that they are comfortable and fit your needs. Binoculars are rated using two numbers, the magnification and the diameter of the main lenses. A pair of binoculars rated 8×30 have a magnification of times 8 and a lens diameter of 30mm. The image you see is determined by these two values. The bigger the magnification, the greater the enlargement and closer the image will appear, but the smaller your overall view will be. The wider the diameter of the lenses, the more light will be let in and the brighter the image will appear. The quality of the lenses is also important and a good quality pair of 8×30 may produce as clear an image as a cheap pair of 8×50. For general use, it is best to choose a pair of 8×30 or 8×40.

LEFT: Viewed through 8× and 10× magnification binoculars, the difference in enlargement of this Chiffchaff is obvious. But 10× magnification binocular are appreciably heavier than 8× pairs, and many people rule them out on this basis alone.

8×   10×

ABOVE RIGHT: In addition to a pair of binoculars, a telescope is also a part of the armoury of the modern birdwatcher.

MAKE YOUR BIRDWATCHING COUNT

Many garden birdwatchers already keep simple records of the birds that they see using their gardens throughout the year. The collection of such information is incredibly useful and, if carried out in a systematic manner, these weekly observations of birds can prove very valuable for researchers. The BTO Garden BirdWatch represents a standardised way of collecting such information and for many people the project is simply an extension of the records they already keep. The structure of Garden BirdWatch, with its weekly recording format and consistent recording effort, greatly increases the scientific value of these observations. By gathering small amounts of simple information from a very large number of gardens, it is possible to answer some complex ecological questions about the relationships between individual bird species and the garden environment. Some of the changes in garden bird populations are likely to reflect changes in the bird populations over a wider area. For example, as farmland habitats become less suitable for seed-eating birds during the winter months we see an increasing reliance on feeding stations within rural and suburban gardens. Although it is not clear how bird numbers reported in gardens relate to their absolute populations, the Garden BirdWatch results can be used alongside other BTO studies to examine how changes in populations of individual species differ across habitats.

THE BTO GARDEN BIRDWATCH METHOD

Launched in 1995, Garden BirdWatch gathers information in a way that makes it possible to measure relative change in the use that birds make of gardens. This approach is similar to that behind other long-running BTO projects and it is particularly suited to large-scale projects covering a wide range of species at many different recording sites. The sheer size of Garden BirdWatch imposes constraints on the type of research questions that can be addressed and the way in which data may be collected. Fortunately, the type of information gathered can readily be coded on forms that can be automatically read by a scanning machine. Some two-thirds of Garden BirdWatchers submit their results on these paper forms, with the remaining third using Garden BirdWatch Online (www.bto.org/gbw) to submit their observations via the Internet. The latter method enables individual Garden BirdWatchers to view all their own data, including those originally submitted on the paper forms. The observations are validated as they are entered and the information is then automatically loaded into the massive Garden BirdWatch Online database. Overnight, various programs run to generate reporting rate graphs, summary tables and scrolling maps showing the location of recent sightings. Alongside the information recorded on birds, other details are gathered on the site at which the recording takes place. These include information about the other features present within the garden recording area, as well as on the nature of surrounding habitats. This information has been used to examine which factors may determine the nature of the bird community visiting gardens at different times of the year.

REPORTING RATES

The vast Garden BirdWatch dataset can be analysed in many different ways. One of the simplest ways of showing how the use of gardens changes over time is by presenting reporting rate graphs (*see graph*). The reporting rate is simply the number of gardens containing the species in a given week, divided by the total number of gardens at which recording was carried out during that week. For example, a reporting rate value of 0.5 would mean that the species was recorded in half (50%) of the gardens that week. More complex analyses can be performed using the count data gathered through Garden BirdWatch or by looking at differences that may exist between regions, habitats or seasons. It is also possible to examine how the use of gardens changes in relation to temperature, snow cover and other weather-related variables.

male Greenfinch

rural ■
suburban ■
urban ■

BTO GARDEN BIRDWATCH REPORTING RATE
(%) THROUGHOUT THE MONTHS OF THE YEAR

LONG-TERM CHANGES

In addition to providing information on the seasonal patterns of garden use, the Garden BirdWatch results can also give us important information on longer-term changes. Increases in Garden BirdWatch reporting rates for species like Goldfinch and Woodpigeon are the result of changes in populations of these species within other habitats. In the case of the Woodpigeon, a change in cropping practices within agricultural land appears to have increased winter survival and has resulted in an increase in its population, particularly in south and east England (see graph). The Goldfinch population in farmland has been recovering from an earlier population decline but the species has also begun to exploit the wider range of supplementary foods and feeding opportunities provided in gardens (see graph).

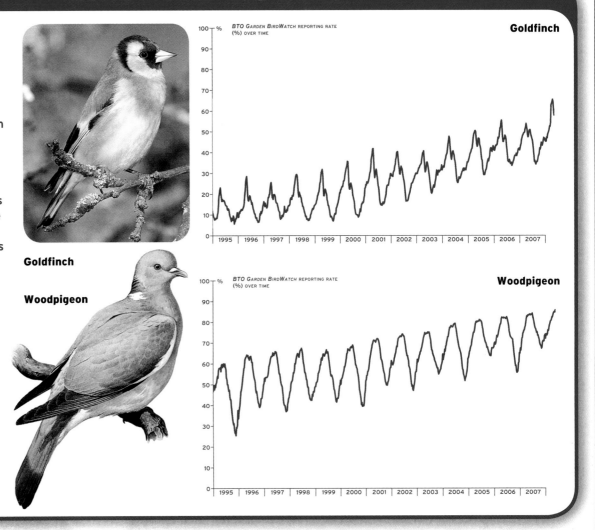

Goldfinch

Woodpigeon

USING THE DATA WISELY

One of the most valuable and exciting aspects of Garden BirdWatch is the range of uses to which the information gathered can be put. In addition to telling us how and why birds use gardens, Garden BirdWatch also provides an insight into what is happening to birds in other habitats and makes a valuable contribution to the BTO's Integrated Population Monitoring approach. Garden BirdWatch would not have been able to fulfil this potential were it not for the generosity and commitment of its Garden BirdWatchers.

Depending on where you live and the particular year, Siskins appear in variable numbers at garden feeders; trends in their use of feeders are detected by BTO surveys.

THE GARDEN BIRD FEEDING SURVEY

The true value of the BTO Garden BirdWatch will increase with time as has been the case with its companion survey, the Garden Bird Feeding Survey (GBFS). GBFS is the longest running annual survey of garden birds anywhere in Europe, having been launched during the 1970/71 winter and currently involving about 250 participants annually. Although the project was originally set up to assess the range of birds using supplementary food provided in gardens throughout the winter months, and to examine feeding preferences, it also provides a very valuable check on how garden bird populations have changed over time. One of the most important aspects of GBFS is that throughout its 33-year run, observers have always used the same recording methodology, allowing comparisons to be made between years. The Garden Bird Feeding Survey involves weekly counts of those bird species using supplementary foods between October and March each year. Gardens included in the project are subdivided into those classed as 'rural' and those classed as 'suburban' in location – currently a roughly equal split across the 250+ sites. For each group of sites a number of different indices can be calculated that reflect the numbers and range of species visiting gardens during a particular winter. One of these is the Peak Count Index. Calculated separately for each species within each garden type, the Peak Count Index is a measure of the maximum number of individuals, of a particular species, seen together, averaged across all of the gardens within the garden type.

GARDENING FOR BIRDS

Birds are arguably the most visible elements of the garden fauna and their presence reinforces our associations with the natural world, adding to the quality of our daily lives. Birds are attracted to our gardens for the resources they offer, such as the prospect of food or nesting opportunities. In turn, these are heavily influenced by the way in which we maintain our gardens. Gardening with birds in mind can add greatly to the avian diversity closest to home.

NESTING OPPORTUNITIES

For most cavity-nesting birds the provision of a suitable nestbox is enough to encourage them to breed in your garden. However, open-nesting species like Greenfinch and Song Thrush are harder to accommodate. The dense cover of woodland or a mature hedgerow needs to be replicated and so planting suitable shrubs and small trees in the garden is a great help. Some birds make their first nesting attempt of the year in an evergreen shrub (when broad-leaved trees are usually still in bud); broad-leaves come into their own later in the season and may be used for later breeding attempts. So establish a mixture of evergreen and broad-leaved trees and shrubs, planting some in small groups and others in the form of a hedge. For rapid evergreen cover (useful in a new garden) try ornamental cypresses (Chamaecyparis), such as C. lawsoniana 'Ellwoodii' (grows to 3m) or C. lawsoniana 'Minima' (rounded shape, grows to 1.5m); Dunnocks, Greenfinches and Goldfinches will use them. Other evergreen options include our native Holly *Ilex aquifolium* (if you have both male and female plants present you will also get flowers for insects and berries for birds) and Yew *Taxus baccata*. Two other useful plants are Ivy *Hedera helix* and the evergreen honeysuckle Lonicera henryi (a semi-hardy Chinese species). Our native Honeysuckle is not evergreen. Among broad-leaves, those that produce thick, thorny cover are best. Try Hawthorn *Crataegus monogyna*, *Berberis* and *Pyracantha*. Sea-buckthorn *Hippophae rhamnoides* is another good shrub, especially in coastal gardens as it is salt-tolerant.

As shrubs go, Hawthorn is a good all-rounder: its dense, spiny twigs and branches provide cover for nesting birds, the flowers are a source of pollen and nectar, its leaves are eaten by moth caterpillars and its fruits feed thrushes and other birds in autumn and winter.

fruits

TREES AND SHRUBS SUITABLE FOR SHELTER AND NESTING COVER

Berberis darwinii ● *Berberis gagnepainii* ● Blackthorn *Prunus spinosa* ● Box *Buxus sempervirens* ● *Cotoneaster lacteus* ● *Cotoneaster horizontalis* ● *Elaegnus ebbingei* ● Hawthorn *Crataegus monogyna* ● Holly *Ilex aquifolium* ● Ivy *Hedera helix* ● *Laurus nobilis* ● Lawson Cypress *Chamaecyparis lawsoniana* ● Leyland Cypress *Cupressocyparis leylandii* ● *Lonicera henryi* ● Privet *Ligustrum vulgare* ● *Prunus laurocerastus* ● *Prunus lusitanica* ● *Pyracantha* spp. ● Sea-buckthorn *Hippophae rhamnoides* ● *Viburnum fragrans* ● *Viburnum tinus* ● Yew *Taxus baccata*.

NESTING REQUIREMENTS OF SOME COMMON GARDEN BIRDS

SPECIES	OPEN/CAVITY NESTER	HEIGHT	LOCATION
Blackbird	Open	<4m	Bush providing thick cover
Blue Tit	Cavity		Tree holes and nestboxes
Chaffinch	Open	1-4m	Canopy or fork of a bush
Coal Tit	Cavity		Tree holes and nestboxes
Dunnock	Open	15cm to 1.5m	Bush providing thick cover
Goldcrest	Open	2-12m	Conifer canopy
Goldfinch	Open	up to 15m	Bush or fork within tree canopy
Great Tit	Cavity		Tree holes and nestboxes
Greenfinch	Open	1-5m	Bush or creeper
Long-tailed Tit	Open		Bush providing thick cover
Mistle Thrush	Open	3-10m	Fork or bow of tree
Robin	Open	<3m	Bush providing thick cover
Song Thrush	Open	<4m	Bush providing thick cover
Spotted Flycatcher	Open/Cavity	<10m	Open cavity or crevice, in a creeper
Starling	Cavity	up to 15m	Cavity or nestbox
Tawny Owl	Cavity		Cavity or nestbox

INFORMATION TAKEN FROM *GARDENING FOR BIRDWATCHERS* BY MIKE TOMS, AND IAN AND BARLEY WILSON

PROVIDING FEEDING OPPORTUNITIES

With many birds, the availability of natural foods in the garden influences, or determines, whether they are present. Invertebrates are important as food for many species while others rely on seeds, nuts and fruits. Unsurprisingly, some trees and shrubs provide better feeding opportunities than others, not only when it comes to seeds, fruits and berries, but also in terms of the invertebrate communities they support. Consequently, small insect-eating birds favour the canopy of Sycamore over Beech – the former supports a greater insect biomass. Birds also show clear preferences over which berries and seeds are taken and when. The key to making your garden attractive to feeding birds is to provide as many feeding opportunities as possible. Providing invertebrate prey hinges on providing the correct environment for the invertebrates themselves; this is covered in the next section. The provision of suitable berries, fruits and seeds is more straightforward since you can select plant species so that you have a range of berries and seeds present throughout as much of the year as possible.

Different fruits (in botanical parlance the term 'fruit' covers fruits, seeds and berries) become available at different times of the year. Some are ephemeral, succulent and avidly eaten by birds as soon as they ripen; others persist for long periods (often because they are relatively unpalatable) and used when all else fails. Providing a mixture of the two is a good idea. The first fruits of autumn appear in June (Wild Cherry), with others (Hawthorn, Blackthorn and Holly) maturing from August onwards. One of the last to appear is Ivy (from November onwards) and these fruits often last through into the following June.

Guelder-rose

Rowan

Ivy

By growing shrubs such as Guelder-rose, Rowan and Ivy in the garden you will provide a feast of berries for birds, from August to December.

WHO EATS WHAT?

A series of studies by Barbara and David Snow highlighted that Blackbirds eat a wide range of fruits (including haws, rosehips, sloes, together with the berries of Holly and Yew). The Song Thrush shows a clear preference for sloes, Yew, Elder and Guelder-rose, and avoids rosehips (presumably because they are too large to swallow). Among the thrushes, only Blackbird, Mistle Thrush and Fieldfare eat rosehips. Late in the season, when berry choice is limited to Holly and Ivy, the preference is for Ivy. Some birds feed on seeds; Goldfinches eat those of Lemon Balm and Teasel, as well as dandelions and thistles. Other species specialise in feeding on larger seeds of our evergreen and broad-leaved trees: Siskins and Coal Tits depend on conifer seeds late in the year, while Bramblings and Chaffinches favour Beech and Lesser Redpolls favour Alder and Birch.

FOOD PLANTS FOR BIRDS

BERRY-BEARING PLANTS

Bird Cherry *Prunus padus* ● Blackberry *Rubus fruticosus* agg. ● Blackthorn *Prunus spinosa* ● *Cotoneaster bullatus* ● *Cotoneaster* 'Cornubia' hybrids ● *Cotonoeaster horizontalis* ● Crab apple *Malus sylvestris* ● Dog Rose *Rosa canina* ● Elder *Sambucus nigra* ● Guelder-rose *Viburnum opulus* ● Hawthorn *Crataegus monogyna* ● Holly *Ilex aquifolium* ● Honeysuckle *Lonicera periclymenum* ● Ivy *Hedera helix* ● Mezereon *Daphne mezereon* ● Midland Hawthorn *Crataegus laevigata* ● Mistletoe *Viscum album* ● Perfoliate Honeysuckle *Lonicera caprifolium* ● Oregon Grape *Mahonia aquifolium* ● *Photinia davidiana* ● *Pyracantha coccinea* ● *Pyracantha rogersiana* ● Rowan *Sorbus aucuparia* ● Sea-buckthorn *Hippophae rhamnoides* ● *Stranvaesia davidiana* ● Wayfaring-tree *Viburnum lantana* ● Whitebeam *Sorbus aria* ● Wild Cherry *Prunus avium* ● Wild Privet *Ligustrum vulgare* ● Wild Service-tree *Sorbus torminalis* ● Yew *Taxus baccata*.

Yew

Privet

Elder

Holly

Crab Apple

SEED-PRODUCING PLANTS

Alder *Alnus glutinosa* ● Beech *Fagus sylvatica* ● Dandelion *Taraxacum officinale* agg. ● Devil's-bit Scabious *Succisa pratensis* ● Field Scabious *Knautia arvensis* ● Greater Knapweed *Centaurea scabiosa* ● Hazel *Corylus avellana* ● Hornbeam *Carpinus betulus* ● Lavender *Lavandula* spp. ● Lemon Balm *Melissa officinalis* ● Silver Birch *Betula pendula* ● Sunflower *Helianthus annuus* ● Teasel *Dipsacus fullonum* ● Thistles *Carduus* spp. and *Cirsium* spp.

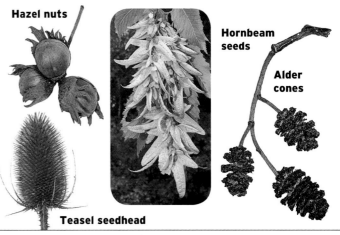

Hazel nuts

Hornbeam seeds

Alder cones

Teasel seedhead

INFORMATION TAKEN FROM *GARDENING FOR BIRDWATCHERS* BY MIKE TOMS, AND IAN AND BARLEY WILSON

NESTBOXES

Many gardens lack the range of nesting opportunities needed by birds for successful breeding and providing artificial nest sites (typically nestboxes) can really help the local bird population. To be successful, artificial nest sites need to be an appropriate design and located in a suitable place. They should also be weatherproof, secure from nest predators and maintained in good condition.

NESTBOXES - THE RIGHT DESIGN

Nestboxes can be divided into three main types: the classic tit box, the open-fronted box and various specialist designs for species like Swift and House Martin. Tit boxes have a complete front panel with a round hole drilled into it. The size of the hole (and the size of the box) determines which species will use it. For smaller birds (e.g. tits, sparrows and Nuthatch) the box itself can be quite small, with a minimum internal floor dimension of 15cm by 12cm, but for larger species (e.g. Starling and Jackdaw) the box needs to be proportionally bigger. For tit and sparrow species, the base of the box should be at least 12cm below the bottom of the entrance hole, so that predatory cats cannot reach the nest contents. In open-fronted boxes, the upper part of the front is cut away, favouring species that like to nest in more open situations (e.g. Robin and Pied Wagtail). An open-fronted design works for Spotted Flycatcher but, because this species likes to be able to see over the edge of the nest while incubating, only a short, 6cm tall, front panel is needed. On the other hand, the Wren favours an open-fronted box with only a 3–4cm gap between the front panel and the roof. Building your own nestbox can be very rewarding and there are a number of good books on the subject that include cutting plans for a range of species. One of the best is *The BTO NestBox Guide by* Chris du Feu. Many people prefer to purchase a nestbox and there is a bewildering array of designs available in pet shops, garden centres and by mail order. Some are wholly inappropriate: for example avoid those that combine a nestbox with a bird table. Look for boxes that are well-made and whose dimensions meet the requirements of the bird species you are hoping to attract.

PLACEMENT AND MAINTENANCE

Although northeast is generally the best direction for a box to face, in reality it makes little difference if it is sheltered from the prevailing wind and rain, and positioned away from strong sunlight. Use available cover to help position the box away from the unwanted attentions of nest predators but ensure that there is a clear flight path in to the box. Make sure that the box is fastened securely to a structure (a tree, wall or post) and that it is in a quiet spot, away from the unwelcome attentions of predators or children. Height is less important than you might imagine but the box is usually best placed at least 2m off the ground. Never place a nestbox adjacent to a bird table or any hanging feeders – the residents may suffer from undue disturbance and can spend too much time trying to defend the area around the box. Boxes should be cleaned out in the autumn (after August), when any chance of further broods is over. This reduces the build-up of nest parasites, like fleas, which remain in the old nest and await the arrival of the next year's brood.

ABOVE: **Satisfied customers – these Great Tit parents are well on the way to rearing a healthy brood of chicks, thanks to the provision of a hole-fronted nestbox.**

FAR LEFT: **A hole-fronted nestbox designed for tits and other songbirds.**

LEFT: **An open-fronted nestbox.**

NESTBOX DIMENSIONS FOR COMMON GARDEN BIRDS

Use wood that is at least 15mm thick and suitable for use outside when constructing a nestbox. In order to prolong the life of the box you may wish to treat the outside of the box with a coat of wood preservative; never treat the inside. Avoid traditional oil-based preservatives but opt instead for one of the non-toxic, water-based preservatives that are now on the market.

SPECIES	TYPE OF BOX	SIZE (mm)	ENTRANCE HOLE (mm)	SITING
BLUE TIT	Hole-fronted box	*base* 150 × 120mm *front* 150 × 175mm	25mm	1-5m off ground, with clear flight path to nest entrance.
GREAT TIT	Hole-fronted box	*base* 150 × 120mm *front* 150 × 175mm	28mm	1-5m off ground, with clear flight path to nest entrance.
COAL TIT	Hole-fronted box	*base* 150 × 120mm *front* 150 × 175mm	25mm	low, will nest higher if no competition for nest site.
HOUSE SPARROW	Hole-fronted box	*base* 150 × 120mm *front* 150 × 175mm	32mm	2-5m off ground, fixed to tree or building.
TREE SPARROW	Hole-fronted box	*base* 150 × 120mm *front* 150 × 175mm	28mm	2-5m off ground, avoid disturbed sites.
NUTHATCH	Hole-fronted box	*base* 150 × 120mm *front* 150 × 175mm	32mm	>3m off ground, with clear flight path to entrance.
STARLING	Hole-fronted box	*base* 150 × 180mm *front* 150 × 250mm	45mm	>2.5m off ground, on tree trunk or building.
JACKDAW	Hole-fronted box	*base* 300 × 300mm *front* 300 × 400mm	150mm	At least 3m off ground, but as high as possible.
ROBIN	Open-fronted box	*base* 150 × 120mm *front* 150 × 100mm		1-3m off ground, well-hidden by thick vegetation.
WREN	Open-fronted box	*base* 150 × 120mm *front* 150 × 140mm		1-3m off ground, well hidden by thick vegetation.
SPOTTED FLYCATCHER	Open-fronted box	*base* 150 × 120mm *front* 150 × 60mm		2-4m off ground, with clear outlook.

INFORMATION TAKEN FROM *GARDENING FOR BIRDWATCHERS* BY MIKE TOMS, AND IAN AND BARLEY WILSON

NEST RECORDING

Since 1939, participants in the British Trust for Ornithology's Nest Record Scheme have gathered information about the breeding success of Britain's birds. Anyone who finds a nest of any species anywhere, including gardens, can take part by making at least two visits to it during the breeding season and recording the number of eggs and chicks inside. Provided due care is taken, visiting a nest will not cause it to fail and the Nest Records Unit provides a set of guidelines specially designed to minimise the risk of disturbance. The data collected are extremely valuable, helping BTO scientists to identify the causes of population declines and investigate the impacts of global climate change. For more information about taking part please write to Nest Record Scheme, BTO, The Nunnery, Thetford, Norfolk IP24 2PU or email nest.records@bto.org.

BTO Nest Record Scheme

WILDLIFE FRIENDLY GARDENS

Although there are many different types of wildlife in an average garden, most of these animals and plants remain unnoticed. But just because you do not see them does not mean that they are unimportant: this diversity of smaller organisms supports the larger creatures that you do notice. Wildlife-friendly gardening focuses on managing the garden for the benefit of all these different creatures. Consider their requirements when gardening and maximise the number of habitats and microhabitats available and you will create your own private nature reserve.

NECTAR AND POLLEN

Many gardeners have favourites in the insect world – bumblebees, butterflies and hoverflies are always popular, for example. Attracting these species is a good starting point for any wildlife-friendly gardener because what suits them also suits a whole host of other insects. Flowers are of particular importance to bumblebees and butterflies, providing pollen, nectar and, on occasions, shelter too.

Bumblebee pollinating a flower of Greater Knapweed.

Some plants are better for bees than others and, as with many other insects, blooms that have been heavily modified by plant breeders (most annual bedding plants and double-flowered cultivars) are largely unsuitable. Many of the plants visited by bumblebees are suited to the bees' relatively long tongues, although tongue length variation exists between bumblebee species, highlighting the need to use a balance of flower forms. Butterflies and moths show similar preferences for particular nectar sources, reflecting variations in their tongue-lengths. In addition to flower form, the position of the flower within the garden will also be important. Most insects favour flowers that are planted in a sunny, sheltered location, so a south-facing border protected from the prevailing wind by a hedgerow or fence is ideal.

LARVAL FOODPLANTS

The larvae (caterpillars) of butterflies and moths also utilise plants – as foodplants. Most eat only a restricted range of plants, a few have only one foodplant species. Egg-laying adults are very selective when choosing the plants on which to deposit their eggs. Butterflies seldom lay eggs on the wrong foodplant (using chemical cues) and the degree of shade, leaf size, microclimate and plant condition also influence where they lay. So wildlife gardeners beware – you may have the right plant but it might be growing in the wrong place. This may explain why it is so hard to attract breeding Commas, Small Tortoiseshells and Peacocks to a garden nettle patch although all three use nettles for egg-laying.

Peacock caterpillars feeding on Common Nettle.

PLANTS FOR NECTAR AND POLLEN

One way to make your garden attractive to a wide range of insects and other invertebrates is to extend the season for pollen and nectar for as long as possible. When planting a new border, or a wildflower area, select plants to create a floral succession from spring to autumn. The following are worth considering:

FLOWERING IN MAY TO JUNE

Aquilegia spp. ● Borage *Borago officinalis* ● *Buddleja globosa* ● Bugle *Ajuga reptans* ● *Campanula* spp. ● *Ceanothus* spp. ● Chives *Allium schoenoprasum* ● Common Bistort *Persicaria bistorta* ● Common Comfrey *Symphytum officinale* ● *Geranium* spp. ● Foxglove *Digitalis purpurea* ● *Hebe* spp. ● Honeywort *Cerinthe major* 'Purpurascens' ● Kidney Vetch *Anthyllis vulneraria* ● Lupin *Lupinus × regalis* ● Red Campion *Silene dioica* ● Red Clover *Trifolium pratense* ● Selfheal *Prunella vulgaris* ● Wallflower *Erysimum cheiri* ● White Clover *Trifolium repens* ● Field Woundwort *Stachys arvensis* ● Yellow-rattle *Rhinanthus minor*.

FAR LEFT: **Bugle**
MIDDLE LEFT: **Common Bistort**
LEFT: **Chives**

FLOWERING IN JULY AND AUGUST

Bramble *Rubus fruticosus* agg. ● burdocks *Arctium* spp. ● Black Horehound *Ballota nigra* ● Butterfly-bush *Buddleja davidii* ● Cat-mint *Nepeta cataria* ● Common Bird's-foot Trefoil *Lotus corniculatus* ● Common Hemp-nettle *Galeopsis tetrahit* ● Common Knapweed *Centaurea nigra* ● Cornflower *Centaurea cyanus* ● *Delphinium elatum* ● Globe-thistles *Echinops exaltatus*, *E. ritro* and *E. bannaticus* ● Great Mullein *Verbascum thapsus* ● Hollyhock *Alcea rosea* ● Iceplant *Sedum spectabile* ● Lavender *Lavandula angustifolia* ● Meadow Clary *Salvia pratensis* ● nasturtiums *Tropaeolum* spp. ● Rosebay Willowherb *Chamerion angustifolium* ● Sainfoin *Onobrychis viciifolia* ● Devil's-bit Scabious *Succisa pratensis* ● thistles *Cirsium* spp. and *Carduus* spp. ● Tufted Vetch *Vicia cracca* ● Viper's-bugloss *Echium vulgare* ● Water Mint *Mentha aquatica* ● Wild Marjoram *Origanum vulgare*.

FAR LEFT: **Cotton Thistle**
MIDDLE LEFT: **Sainfoin**
LEFT: **Devil's-bit Scabious**

OTHER INVERTEBRATES

When it comes to other invertebrates, there are four ways in which you can make a positive difference for a wide range of species. **1.** Avoid the use of chemicals in the garden. Insecticides, molluscicides and herbicides inevitably have a knock-on impact on non-target species and are seldom as selective as manufacturers might have us believe; they are best avoided so use other (hands-on and organic) methods to tackle pests and diseases. **2.** Maximise invertebrate biodiversity by increasing the number of microhabits and food plants in your garden; whole communities may depend on the decaying wood of your shed or the thatch in your lawn. **3.** Provide shelter through the provision of log-piles and by leaving dead flower heads in situ through the winter months. The key is not to be too tidy-minded. **4.** Provide breeding sites, for example by investing in a mason bee nesting kit or by leaving parts of the garden to go 'wild'.

Centipedes, and many other garden invertebrates, benefit from people shedding the tidy-minded approach to gardening, and leaving log piles and other places that serve as daytime refuges.

NATIVE OR NON-NATIVE?

Debate continues as to whether non-native plants have a place in the wildlife-friendly garden. The answer is not clear-cut, because although native plants are part of our natural wildlife heritage, many non-native plants have tremendous wildlife value too. Most gardens contain a mix of the two and the degree to which a particular plant (native or otherwise) is useful to an organism will depend on how it is used by the organism in question. If structural features are important, as shelter for example, then it matters little if the plant is native or non-native. However, sometimes a plant's chemistry makes it unsuitable as a foodplant for caterpillars, or its flower structure may exclude pollination and nectar collection by native insects; this is more likely to be the case with non-native than native plants. And the invasive qualities of some non-native plants (Rhododendron and Japanese Knotweed, for example) mean they are anything but wildlife friendly. Despite this, non-native plants can have a role in a wildlife-friendly garden, particularly when used to extend the nectar season. But don't neglect our native species – after all, they are wildlife too.

GARDEN BUTTERFLY LARVAL FOODPLANTS

An awareness of the foodplants of butterfly larvae will help the wildlife-friendly gardener in their efforts to benefit all stages in the life-cycles of these charming insects. While most butterfly larvae eat only native plant species, a few relish cultivated and ornamental non-native ones.

SPECIES	FOODPLANTS
LARGE WHITE *Pieris brassicae*	Brassicas; Nasturtium *Tropaeolum majus*
SMALL WHITE *Pieris rapae*	Brassicas; Nasturtium *Tropaeolum majus*; Hedge Mustard *Sisymbrium officinale*
GREEN-VEINED WHITE *Pieris napi*	Garlic Mustard *Alliaria petiolata*; Cuckooflower *Cardamine pratensis*
ORANGE-TIP *Anthocharis cardamines*	Cuckooflower *Cardamine pratensis*; Garlic Mustard *Alliaria petiolata*; Honesty *Lunaria annua*; Dame's-violet *Hesperis matronalis*
BRIMSTONE *Gonepteryx rhamni*	Buckthorn *Rhamnus cathartica*; Alder Buckthorn *Frangula alnus*
SMALL TORTOISESHELL *Nymphalis urticae*	Common Nettle *Urtica dioica*; Small Nettle *Urtica urens*
PAINTED LADY *Vanessa cardui*	Thistles *Cirsium* spp. and *Carduus* spp.; mallows *Malva* spp.
RED ADMIRAL *Vanessa atalanta*	Common Nettle *Urtica dioica*; Small Nettle *Urtica urens*; Pellitory-of-the-wall *Parietaria judaica*; Hop *Humulus lupulus*
PEACOCK *Nymphalis io*	Common Nettle *Urtica dioica*
COMMA *Polygonia c-album*	Common Nettle *Urtica dioica*; Hop *Humulus lupulus*; currants *Ribes* spp.
SPECKLED WOOD *Pararge aegeria*	Cock's-foot *Dactylis glomerata*; Yorkshire-fog *Holcus lanatus*
GATEKEEPER *Pyronia tithonus*	Various grasses, including bents *Agrostis* spp. and fescues *Festuca* spp.
MEADOW BROWN *Maniola jurtina*	Various grasses, including bents *Agrostis* spp. and fescues *Festuca* spp.
SMALL COPPER *Lycaena phlaeas*	Common Sorrel *Rumex acetosa*; Sheep's Sorrel *Rumex acetosella*
COMMON BLUE *Polyommatus icarus*	Common Bird's-foot Trefoil *Lotus corniculatus*
HOLLY BLUE *Celastrina argiolus*	Holly *Ilex aquifolium*; Ivy *Hedera helix*; Spindle *Euonymus europaeus*
SMALL SKIPPER *Thymelicus sylvestris*	Yorkshire Fog *Holcus lanatus*
LARGE SKIPPER *Ochlodes venatus*	Cock's-foot *Dactylis glomerata*

INFORMATION TAKEN FROM *GARDENING FOR BIRDWATCHERS* BY MIKE TOMS, AND IAN AND BARLEY WILSON

PONDS FOR WILDLIFE

A pond is a stunning feature in any garden and adds instant wildlife appeal: the addition of one to a new garden will make an amazing difference to the range of visiting and resident species. Birds will come to drink and bathe and insects including water beetles and dragonflies will soon colonise. And if you are lucky, frogs and other amphibians will breed in spring.

Pond in the BTO's garden at The Nunnery, Thetford.

POND CONSTRUCTION

A pond is one of the easiest features to establish in the garden although while digging it out you may come to doubt this. Size is important: generally speaking the bigger (and deeper) the better, but even a small pond can make a difference to the wildlife value of a garden. Bigger ponds are better because the water temperature remains more stable, they can support more species and are less likely to suffer from algal blooms. A good size would be 3m long by 1.5m wide and about 0.75m deep at the deepest point. Ensure that there are some shallow margins and gently sloping sides, allowing access to wildlife that may come to bathe or drink from the pond. Try to make the pond blend into the landscape; not only does this make the pond more appealing visually but it also allows wildlife to enter and leave the pond without attracting the unwanted attentions of predatory cats or birds. The pond itself can be constructed from a moulded plastic or fibreglass shell, clay, concrete or with a flexible liner. Once the pond has been established and planted with suitable plants, you will soon find that wildlife moves in of its own accord. If you have managed to get the right balance of plants and nutrients, the pond should need little in the way of day-to-day management. If it needs topping up in the summer, ideally use rainwater from a water butt. Try to locate the pond where it receives shade from trees but not directly beneath them (to avoid autumn leaf fall); remove any leaves that have accumulated in early winter. If you want to encourage invertebrates and amphibians, do not introduce fish to the pond.

PLANTS FOR A WILDLIFE POND

OPEN WATER PLANTS
Amphibious Bistort *Persicaria amphibia* ● Fringed Water-lily *Nymphoides peltata* ● Frogbit *Hydrocharis morsus-ranae* ● Yellow Water-lily *Nuphar lutea* ● Water-soldier *Stratiotes aloides* ● White Water-lily *Nymphaea alba* ● Common Water-crowfoot *Ranunculus aquatilis* ● Water-violet *Hottonia palustris*.

MARGINAL PLANTS
Arrowhead *Sagittaria sagittifolia* ● Bogbean *Menyanthes trifoliate* ● Brooklime *Veronica beccabunga* ● Yellow Iris *Iris pseudacorus* ● Marsh-marigold *Caltha palustris* ● Hard Rush *Juncus inflexus* ● Cyperus Sedge *Cyperus pseudocyperus* ● Remote Sedge *Carex remota* ● Lesser Spearwort *Ranunculus flammula* ● Water Forget-me-not *Myosotis scorpioides* ● Water-plantain *Alisma plantago-aquatica* ● Blue Water-speedwell *Veronica anagallis-aquatica* ● Water-cress *Rorippa nasturtium-aquaticum* ● Purple Loosestrife *Lythrum salicaria*.

ABOVE LEFT:
Fringed Water-lily
ABOVE RIGHT:
Yellow Iris
LEFT:
Cuckooflower
RIGHT:
Curled Pondweed

PLANTS FOR DAMP EDGES
Meadowsweet *Filipendula ulmaria* ● Marsh Woundwort *Stachys palustris* ● Water Avens *Geum rivale* ● Ragged Robin *Lychnis flos-cuculi* ● Greater Bird's-foot Trefoil *Lotus pedunculatus* ● Red Campion *Silene dioica* ● Bugle *Ajuga reptans* ● Cuckooflower *Cardamine pratensis* ● Hemp-agrimony *Eupatorium cannabinum* ● Primrose *Primula vulgaris* ● Meadow Vetchling *Lathyrus pratensis*.

OXYGENATORS
Curled Pondweed *Potamogeton crispus* ● Common Water-starwort *Callitriche stagnalis* ● Rigid Hornwort *Ceratophyllum demersum* ● Spiked Water Milfoil *Myriophyllum spicatum*.

POND LIFE

Freshwater invertebrates are quick to colonise a new garden pond although the diversity present will depend on the age of the pond, its size and proximity to other bodies of water. Within a couple of years even the most unpromising of sites will have a good selection of the creatures shown here.

SOUTHERN HAWKER
Aeshna cyanea LENGTH 70mm
Active predator with blue and green body markings. Predatory nymph is aquatic.

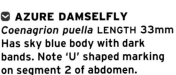

BROAD-BODIED CHASER ▶
Libellula depressa
LENGTH 43mm
Has flattened body, blue in male, orange and yellow in female. Nymph is aquatic.

Southern Hawker empty nymphal skin.

female

male

COMMON DARTER *Sympetrum striolatum* LENGTH 36mm
Often perches near pond, darting after prey. Predatory nymph is aquatic.

COMMON BLUE DAMSELFLY
Enallagma cyathigerum
LENGTH 32mm
Has sky blue body with dark bands. Note mushroom-cloud marking on segment 2 of abdomen.

BLUE-TAILED DAMSELFLY
Ishnura elegans LENGTH 32mm
Has mainly black body with sky blue band near tip of abdomen.

AZURE DAMSELFLY
Coenagrion puella LENGTH 33mm
Has sky blue body with dark bands. Note 'U' shaped marking on segment 2 of abdomen.

RED-EYED DAMSELFLY
Erythromma najus LENGTH 35mm
Has pale blue body with dark bands and lines, and distinctive bright red eyes.

GREAT DIVING BEETLE
Dytiscus marginalis LENGTH 30mm
Impressive, predatory beetle with painful bite. Aquatic larva is also predatory.

POND SKATER
Gerris lacustris LENGTH 10mm
Skates over water surface on tips of legs. Feeds on insects trapped in surface film.

WATER BOATMAN
Notonecta glauca LENGTH 14mm
Hangs upside down at water surface. Feeds on insects trapped in surface film.

LESSER WATER BOATMAN
Corixa punctata LENGTH 10mm
Swims actively (right way up) using fringed hind legs as paddles. Feeds on detritus.

RAMSHORN *Planorbis planorbis* LENGTH 12mm
Spirally coiled shell is smaller and more flattened than Great Ramshorn.

GREAT POND SNAIL
Lymnaea stagnalis LENGTH 45mm
Glides over water plants and under surface film. Visits surface to breathe.

GREAT RAMSHORN ▶
Planorbis corneus
LENGTH 25mm
Large spiral-shelled water snail that tolerates relatively low oxygen levels.

WHIRLYGIG BEETLE ▶
Gyrinus natator
LENGTH 5mm
Seen whirling around on water surface, often in groups. Sensitive to disturbance.

SHRUBS AND TREES

Environmentally aware gardeners know that native trees and shrubs are 'wildlife' too in the grand scheme of things. Encourage and grow them and you will benefit native insects and other invertebrates that depend on them for food.

GREY WILLOW

Salix cinerea HEIGHT TO 6M

Common and widespread native willow that grows to form a variable, much divided shrub, or sometimes a small tree. It has downy shoots and its oblong leaves are glossy above but often develop rusty hairs on the undersurface by autumn. Male catkins are ovoid and yellow while female catkins are similar but greener; they produce plumed seeds. The leaves of Grey Willow are the foodplant for the larvae of many moth species, as well as numerous other insects.

GOAT WILLOW

Salix caprea HEIGHT TO 12M

Sometimes referred to as Sallow, Goat Willow forms a dense shrub or a small tree with stiff shoots. Its oval leaves are up to 12cm long; they have a twisted point at the tip and are woolly below. Male and female catkins appear on separate trees before the leaves and are 2.5cm long with silky hairs before opening. Male catkins are dusted with yellow pollen. Goat Willow is native and is extremely important to wildlife, and in particular to the larvae of moth species.

Grey Willow

Goat Willow

Common Alder

SILVER BIRCH

Betula pendula HEIGHT TO 26M

Familiar native tree. Its basal bark is fissured and forms rectangular plates; higher up the trunk it is smooth and silvery-white. Silver Birch branches are ascending but the twigs and shoots are pendulous, giving mature trees a weeping habit. The triangular, toothed leaves are up to 7cm long and turn golden in autumn. Male catkins are yellow and pendulous while female catkins are greenish. Its leaves provide food for many insects and numerous fungi are associated with its roots and timber.

COMMON ALDER

Alnus glutinosa HEIGHT TO 25M

Widespread, native tree that prefers to grow with its 'feet in water'. Tolerates heavy cutting back and it is often multi-stemmed as a result. The brownish bark is fissured into squarish plates. Young trees have ascending branches and young twigs are sticky, with stalked buds. Alder leaves are stalked, to 10cm long, rounded with a notched apex. The purplish male catkins appear in winter in bunches of 2–3; female catkins are cone-like, reddish first, ripening green by summer.

HAZEL

Corylus avellana HEIGHT TO 6M

Familiar native shrub or small tree that responds well to coppicing; as a result it is often multi-stemmed. Its shiny, smooth bark peels into papery strips and its twigs have stiff hairs and smooth, oval buds. Hazel leaves are rounded, to 10cm long with a heart-shaped base and pointed tip; the margins are double-toothed. Male catkins are 8cm long, pendulous and yellow while female flowers are red and tiny. Hard-shelled hazelnuts appear in late summer and provide a feast for small mammals.

Hazel nuts

HORNBEAM

Carpinus betulus HEIGHT TO 30M

Impressive native tree with silvery-grey, fissured bark and a gnarled and twisted bole in large specimens. Its branches are ascending and twisted, and its twigs are hairy. Hornbeam leaves are oval and pointed with a rounded base and double-toothed margin; they have 15 pairs of veins. Male catkins are yellowish green with red scales. Hornbeam fruits are winged and borne in clusters; they are notoriously hard-cased and Hawfinches are the only birds with bills powerful enough to crack them.

fruits

BEECH

Fagus sylvatica HEIGHT TO 40M

Popular native tree with smooth, grey bark and a domed crown in large specimens. Its branches are ascending and the buds are reddish, up to 2cm long, smooth and pointed. Beech leaves are 10cm long, oval, pointed, with wavy margins. Male flowers are pendent and clustered while female flowers are paired with brownish bracts. The fruits ('mast') are shiny 3-sided nuts, 1.8cm long, enclosed in a prickly case. Fallen mast is eaten by winter flocks of Chaffinches and Bramblings.

Beech mast

SESSILE OAK

Quercus petraea HEIGHT TO 40M

Sturdy, domed tree and the western and upland counterpart of Pedunculate Oak; identified by studying the leaves and acorns. Its bark is grey-brown and fissured and its buds have long white hairs. Sessile Oak leaves are lobed, dark green with hairs below on the veins; they are borne on stalks, 1–2.5cm long. The flowers are catkins and the acorns are stalkless, sitting directly on the twigs in small clusters. It grows well on poor soils and areas of high rainfall.

Sessile Oak acorns

WYCH ELM

Ulmus glabra HEIGHT TO 40M

Spreading tree that is impressive in maturity; sadly, large trees are seldom seen these days thanks to the ravages of Dutch Elm Disease. The bark is cracked and becomes ridged with age, and the young twigs have stiff hairs. Wych Elm leaves are oval, up to 18cm long, and have a tapering tip; their unequal base extends beyond the petiole junction. The fruits are papery and 2cm long. These days, Wych Elm is seen mainly as a hedgerow shrub.

English Elm

LONDON PLANE

Platanus × hispanica HEIGHT TO 44M

Impressive and stately hybrid tree with a tall trunk and a spreading crown; it is widely planted in towns and cities for ornament and shade. The bark is grey-brown and flakes distinctively into patches, and the branches are typically rather tangled. London Plane leaves are 5-lobed, palmate and up to 30cm long. Its flowers are rounded and borne in hanging clusters, and its greenish, spherical fruits (also in hanging clusters) have spiky hairs.

fruits

PEDUNCULATE OAK

Quercus robur HEIGHT TO 36M

Sometimes known as 'English Oak' this spreading, native tree has an iconic conservation status. Its bark is grey, thick and fissured with age. Pedunculate Oak leaves are deeply lobed and borne on very short stalks (5mm or less). The flowers are catkins and fruits are familiar acorns, produced in groups of 1–3, with long stalks and scaly cups. It is immensely important to wildlife: the leaves are food for insects, which in turn are eaten by numerous birds.

HORSE-CHESTNUT

Aesculus hippocastanum HEIGHT TO 25M

Spreading tree with a domed crown. Its bark is greyish brown and flaking, and its branches snap easily. Winter buds are shiny brown, sticky, and have shield-shaped leaf scars. Horse-chestnut leaves are palmate, with up to 7 leaflets, each 25cm long. The 5-petalled, pink-spotted white flowers are borne in spires, up to 30cm tall. The fruits are spiny cased and rounded, containing a round seed ('conker'). A Balkans native, long-established here.

ENGLISH ELM

Ulmus procera HEIGHT TO 36M

Native tree that is tall, with a domed crown in maturity; large trees are rare due to Dutch Elm Disease. The bark is grooved and has squarish plates and the twigs are reddish and hairy. English Elm leaves are rough, rounded to oval in outline with unequal bases that do extend beyond the petiole junction. The fruits are papery, up to 1.5cm long, and short-stalked. English Elm suckers freely and sometimes forms extensive stands as a hedgerow shrub.

SYCAMORE

Acer pseudoplatanus HEIGHT TO 35M

Vigorous tree, introduced but now widely naturalised. Its bark is greyish, fissured and flaking and it has grey-green twigs and reddish buds. Sycamore leaves have 5 toothed lobes and are 15cm long. The flowers are borne in pendulous, yellow clusters and its fruits' paired wings spread acutely, curving in slightly towards the tip. Being invasive, it is often seen as a problem; in its favour, the aphids it supports are food for warblers.

fruit

SHRUBS AND TREES

Plant native shrubs and smaller trees in the garden and you can create your own version of a native hedgerow. By contrast, some introduced, alien shrubs should be treated with caution and a few should perhaps come with an 'environmental health warning'.

FIELD MAPLE
Acer campestre HEIGHT TO 26M

Native tree with a rounded crown and twisted bole. Its bark is corky, grey-brown and fissured. The branches are much divided and its shoots are hairy and sometimes winged. Field Maple leaves are 3-lobed and 10–12cm long; they turn yellow in autumn, making the tree easy to spot at that time. The yellowish flowers are borne in erect clusters and the fruits are reddish and winged. Field Maple is common in woods and hedgerows and does well on chalky soils.

Field Maple fruit

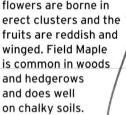

ASH
Fraxinus excelsior HEIGHT TO 40M

Common native tree that is frequent in woods and hedgerows, doing best on chalky or base-rich soils. It has an open crown and its bark is grey, becoming fissured with age. The grey twigs are flattened at the nodes and have conical black buds. Ash leaves are 30cm long and pinnate, with 7–13 narrow, toothed leaflets. Its flowers are small, purple and clustered, and the fruits are single-winged 'keys' that are borne in bunches.

Ash

fruits

HOLLY
Ilex aquifolium HEIGHT TO 15M

Distinctive and familiar evergreen shrub or tree. Its bark is silver-grey, and it becomes fissured with age. The branches sweep downwards but the tips turn up. Holly leaves are leathery and variably wavy with spiny margins. The flowers are 6mm across, 4-petalled, and clustered in the leaf axils; males and females grow on different trees. Its fruits are the familiar red berries, which are eagerly consumed by thrushes in winter. Holly is widespread in hedgerows and woodland.

ROWAN
Sorbus aucuparia HEIGHT TO 20M

Open, deciduous tree that is a common native species, but also widely planted in parks and gardens. Its bark is silvery grey and smooth and the ascending branches have purple-tinged twigs and hairy buds. Rowan leaves are pinnate, with 5–8 pairs of ovate, toothed leaflets, each to 6cm long. The flowers have 5 white petals and are borne in dense heads. Its fruits are rounded, scarlet, and borne in clusters; they are eaten by thrushes and Waxwings in winter.

COMMON WHITEBEAM
Sorbus aria HEIGHT TO 25M

Deciduous tree or spreading shrub that is native in southern England on chalky soils, but also widely planted in urban areas. It has smooth, grey bark and spreading branches with twigs that are brown above, green below. Common Whitebeam leaves are oval, up to 12cm long, toothed and very hairy below. The white flowers are borne in clusters and the ovoid fruits are red. They are eaten by birds such as thrushes in autumn and winter.

ABOVE: **flowers**
LEFT: **fruits**

WILD CRAB
Malus sylvestris HEIGHT TO 10M

fruit

Slender native tree that is locally common in hedgerows and woods. Its deep brown bark cracks into oblong plates and its branches are often spiny. Wild Crab leaves are 10cm long, oval and toothed. Its flowers are 5-petalled, 4cm across and white, sometimes pink-tinged. Wild Crab fruits are miniature apples, 4cm across, rounded and yellowish green; they are hard and too sour for most human palates; deer and birds eat fallen fruits as they start to decay.

WILD CHERRY

Prunus avium HEIGHT TO 30M

Native tree with domed crown, found in woods and hedgerows. It is at its most attractive in spring, when in bloom. The reddish-brown bark is shiny, with distinctive circular lines; it peels horizontally into papery strips. The spreading branches have reddish twigs. Wild Cherry leaves are 15cm long, oval and toothed. The flowers are white, 5-petalled, and borne in clusters of 2–6. Its fruits are familiar cherries, up to 2cm across, rounded, and ripening dark purple, sometimes yellowish.

ABOVE: **fruits**
LEFT: **flowers**

CHERRY PLUM

Prunus cerasifera HEIGHT TO 8M

Introduced shrub; widely planted and often naturalised. The dark brown bark is pitted with white lenticels, and the spiny branches have glossy-green twigs. Cherry Plum leaves are 5–7cm long, oval and toothed; some cultivars have red, not green, leaves. The stalked flowers are usually white, but pink in some cultivars; they appear in February and are often confused with those of Blackthorn. Fruits 3.5cm long, red or yellow.

LEFT: **fruit**
BELOW: **flowers**

COMMON HAWTHORN

Crataegus monogyna
HEIGHT TO 15M

Spreading and dense native tree or shrub that is a common element in many hedgerows, especially on chalky soils. Its bark is fissured with vertical grooves, and its branches are densely packed, with sharp spines. Hawthorn leaves are 4cm long, deeply lobed, and have teeth near the apex. The white flowers are 15mm across, and borne in flat-topped clusters of 10–18; they appear in May. The fruits (known as 'haws') are rounded and red, and eaten by birds and small mammals.

fruits

Common
Hawthorn
leaf

RHODODENDRON

Rhododendron ponticum
HEIGHT TO 6M

Ornamental shrub, introduced from Asia by Victorian horticulturists. The leaves are shiny, leathery, elliptical in outline and dark green while the flowers are pinkish red, bell-shaped and 4–6cm long; they are borne in clusters and appear in May and June. In the context of a garden, Rhododendron may not look out of place. But on acid soils it spreads to the countryside and is so invasive that it swamps native vegetation, requiring its removal.

BLACKTHORN

Prunus spinosa HEIGHT TO 6M

Densely branched native shrub and a common hedgerow component. Its bark is blackish brown and the branches are spreading, with spiny twigs. Blackthorn leaves are oval, toothed and 3–4cm long. Its white flowers are 5-petalled and 1.5cm across; they are produced prolifically, mostly in March and early April, several weeks after those of superficially similar and equally common Cherry Plum. Blackthorn fruits (Sloes) are 1.5cm long, ovoid, and blue-black.

ABOVE: **fruit**
LEFT: **flowers**

CHERRY LAUREL

Prunus laurocerasus HEIGHT TO 8M

Evergreen shrub or small tree that is introduced to Britain but widely planted and sometimes naturalised. Its dark grey-brown bark is pitted with lenticels and the branches are dense, with pale-green twigs. Cherry Laurel leaves are leathery, up to 20cm long and oblong. Its flowers are white, fragrant, and borne in erect spikes up to 12cm long. The rounded fruits are borne in spikes; they are green at first, turning red, then ripening blackish purple.

flowers   **fruits**

LIME

Tilia × europaea
HEIGHT TO 46M

Familiar urban and parkland tree that is a hybrid between native Small-leaved and Large-leaved limes. It suckers freely, its grey-brown bark is ridged, and the arching branches bear green twigs. Lime leaves are 10cm long and oval with a heart-shaped base; they have hairs in the vein axils below. The flowers are yellowish, 5-petalled, and clustered, with a greenish bract. The fruits are hard and rounded. Lime supports large numbers of aphids, which in turn are food for birds.

fruits

BUTTERFLY-BUSH

Buddleia davidii HEIGHT TO 4M

Introduced shrub that is grown in gardens, but often becomes naturalised in the countryside. Its bark is grey-brown and the branches are dense and arching. Butterfly-bush leaves are long and narrow, darker above than below. Its flowers are pinkish purple, 4-lobed, 2–4mm across, and borne in long spikes. Butterfly-bush flowers are extremely attractive to insects – especially butterflies – but its invasive habits mean it has to be controlled where it out-competes native plants.

SHRUBS AND WILDFLOWERS

Many native shrubs and wildflowers produce edible fruits and their berries and seeds provide a feeding bonanza for birds, small mammals and invertebrates in the autumn. Grow or encourage these plants in the garden and you will be providing a ready-made wildlife larder.

Wild Privet fruit

DOGWOOD

Cornus sanguinea HEIGHT TO 4M

Native shrub or small tree that thrives in hedgerows on chalky soils and is often planted elsewhere. Its bark is grey and smooth and the twigs are distinctively dark red in winter. Dogwood leaves are oval with untoothed margins and 3–4 pairs of prominent veins. The flowers are small and white, and are borne in large terminal clusters. Its fruits are blackish, rounded berries that are also borne in clusters. They are popular with birds in autumn.

ABOVE: **fruits;** BELOW: **flowers**

flowers

WILD PRIVET

Ligustrum vulgare HEIGHT TO 5M

Branched, semi-evergreen shrub that is a common native component of hedgerows and scrub on chalky soils; it is also widely planted elsewhere. The reddish-brown bark has distinct gashes (as if cut by a knife) and the twigs are downy. Privet leaves are shiny, untoothed, oval, and opposite. The creamy white flowers are 4–5mm across, fragrant and 4-petalled; they are borne in terminal spikes. Its fruits are shiny, black and clustered and are popular with birds in autumn.

ELDER

Sambucus nigra HEIGHT TO 10M

Untidy deciduous native shrub or small tree that is common in scrub and hedgerows. Its bark is grey-brown and furrowed, becoming corky and lichen-covered with age. The branches are twisted, and have white central pith. Elder leaves comprise 5–7 pairs of ovate, toothed leaflets, each 12cm long. The white flowers have a sickly sweet scent and are borne in flat-topped clusters. Elder fruits are rounded, shiny black berries, borne in pendulous heads; they are popular with birds in autumn.

ABOVE: **Elder flowers**
RIGHT: **Elder fruit**

GUELDER-ROSE

Viburnum opulus HEIGHT TO 4M

Spreading native tree that is found mainly on calcareous soils. Its bark is reddish brown and its sinuous branches bear twigs that are smooth, angular and greyish. Guelder-rose leaves are up to 8cm long, and have 3–5 irregularly toothed lobes. The white flowers are borne in flat heads with showy, outer flowers and smaller, inner ones. Guelder-rose fruits are rounded, translucent red berries that are borne in clusters.

LEFT: **fruit** BELOW: **flowers**

WAYFARING-TREE

Viburnum lantana HEIGHT TO 6M

Native shrub or small, spreading tree that grows mainly on calcareous soils but is also planted elsewhere. The bark is brown and the branches bear rounded, greyish, hairy twigs. Wayfaring-tree leaves are oval, with toothed margins. The white flowers are up to 8mm across, and are borne in rounded heads, up to 10cm across. Wayfaring-tree fruits are oval berries about 8mm long; they ripen red to black and are popular with birds.

LEFT: **fruit** BELOW: **flowers**

HONEYSUCKLE

Lonicera periclymenum HEIGHT TO 5M

Woody climber that twines clockwise through shrubs and trees. Common in woods and hedgerows, and often planted in gardens, mainly for its scented flowers. These are 3–5cm long, trumpet-shaped, 2-lipped and creamy yellow to white; they are borne in whorled heads and appear June–August. Honeysuckle fruits are red berries; borne in clusters, eaten by birds and small mammals. Leaves are grey-green, oval and opposite.

RIGHT: **fruit** BELOW: **flowers**

TRAVELLER'S-JOY

Clematis vitalba LENGTH TO 20M

Scrambling native perennial; favours chalky soils and common in southern Britain. It is found in hedgerows and scrub, sometimes smothering the shrubs it grows through by autumn. Its flowers are creamy white and have prominent stamens; they are borne in clusters and appear July–August. Traveller's-joy fruits comprise clusters of seeds with woolly, whitish plumes, hence the plant's alternative name of Old Man's Beard. Its leaves have 3–5 leaflets.

LEFT: **flowers**
BELOW: **fruits**

IVY

Hedera helix
HEIGHT TO 7M

Familiar plant that grows up trees and carpets the ground. Its yellowish-green flowers are borne in globular heads and appear September–November. Ivy fruits are berries that ripen purplish black and its leaves are glossy, dark green and 3- or 5-lobed with paler veins. Ivy's importance to wildlife cannot be overstated: its flowers provide nectar for insects, its fruits are food for birds in late winter and its leaf cover provides safe roosting sites for birds and hibernating insects.

fruits

BLACK BRYONY

Tamus communis HEIGHT TO 3M

Twining perennial that is widespread and common in southern Britain and grows in hedgerows and scrub. It is similar to unrelated White Bryony (see below): note the different leaf shape and lack of tendrils. Black Bryony flowers are tiny, yellowish green and 6-petalled; they are borne on separate-sex plants and appear May–August. The plant is easiest to spot when its luscious looking red berries ripen in autumn hedgerows. Black Bryony leaves are heart-shaped, glossy and netted.

ABOVE: **fruits**
RIGHT: **flowers**

WHITE BRYONY

Bryonia dioica HEIGHT TO 4M

Climbing hedgerow perennial whose progress is aided by long, unbranched tendrils cf. Black Bryony (see above), which lacks tendrils and has a different leaf shape. White Bryony is common in hedgerows in England, but scarce elsewhere. Its flowers are greenish, 5-parted and are borne on separate-sex plants; they arise from the leaf axils and appear May–August. White Bryony fruits are red, shiny berries that ripen in autumn. Its leaves are 4–7cm across and divided into 5 lobes.

LEFT: **fruits**
ABOVE: **flowers**

COMMON NETTLE

Urtica dioica HEIGHT TO 1M

Usually present in the garden whether you like it or not, this is the familiar stinging nettle. Its flowers are pendulous catkins, borne on separate sex plants (June–October) and its leaves are oval, pointed-tipped and toothed. Common Nettle is widespread and common throughout and thrives on nitrogen-enriched and disturbed soils. Its leaves are food for the larvae of butterflies such as the Small Tortoiseshell, and a few moth larvae too, and its seeds are eaten by finches.

REDSHANK

Persicaria maculosa
HEIGHT TO 60CM

Upright or sprawling hairless plant with much-branched stems whose reddish colour gives the plant its name. It favours disturbed ground and turns up in vegetable plots in the garden; in the countryside, it is commonly found around arable field margins and beside tracks. Redshank flowers are pink and borne in terminal spikes, appearing June–October. Its fruits are nut-like and are eaten by birds, and its leaves are narrow and oval, typically with a dark central mark.

BLACK-BINDWEED

Fallopia convolvulus HEIGHT TO 1M

Extremely common, clockwise-twining annual that both trails on the ground and climbs among wayside plants. It is widespread and common in cultivated soil and on waste ground, and its frequent presence in garden borders is not always welcome. Black-bindweed flowers are greenish and rather dock-like; they are borne in loose spikes that arise from leaf axils and appear July–October. Its fruits are nut-like and blackish and the leaves are arrow-shaped and borne on angular stems.

BROAD-LEAVED DOCK

Rumex obtusifolius HEIGHT TO 1M

Familiar, robust plant that favours disturbed ground and commonly grows in neglected parts of the garden; in the countryside it favours field margins and disturbed grassland. Broad-leaved Dock flowers are borne in loose spikes that are leafy at the base, and appear June–August. Its fruits have prominent teeth and the leaves are broadly oval, heart-shaped at the base and up to 25cm long. The leaves are eaten by a few moth larvae and other insects, and its seeds by birds.

Grow vegetables in the garden and you will encourage invasive native plant species whether you like it or not. But set aside a shady, grassy part of the garden for native flowers and you can create something positive: your own 'woodland ride', at its most colourful in spring.

FAT-HEN

Chenopodium album HEIGHT TO 1M

Upright, branched plant that often has a mealy appearance. It thrives in disturbed ground and so often turns up in the garden vegetable patch; in the countryside, it favours arable fields and waste ground. Fat-hen flowers are whitish green and borne in leafy spikes, appearing June–October. Its fruits are rounded and surrounded by a ring of 5 sepals, and the leaves are green, and matt-looking due to the mealy coating; they vary from oval to diamond-shaped.

RED GOOSEFOOT

Chenopodium rubrum HEIGHT TO 60CM

Variable upright plant whose stems often turn red in old specimens, or if water is at a premium in the summer months. It is commonest in southern England and thrives in manure-enriched soils; unsurprisingly, it often appears in the garden vegetable patch. Red Goosefoot flowers are small and numerous; they are borne in upright, leafy spikes and appear July–October. The fruits are rounded and enclosed by 2–4 sepals and the leaves are shiny, diamond-shaped and toothed.

GREATER STITCHWORT

Stellaria holostea HEIGHT TO 50CM

Familiar wayside plant of hedgerows, woodland rides and roadside verges; it sometimes grows around the margins of rural gardens too. Unlike the superficially similar, smaller flowered Lesser Stitchwort (see below), Greater has rough-edged stems. Its flowers are white, up to 2cm across with 5 notched petals; they are borne on slender stems and appear April–June. The narrow leaves are fresh green, rough-edged, extremely grass-like and easily overlooked.

COMMON CHICKWEED

Stellaria media HEIGHT TO 30CM

Sprawling, sometimes prostrate plant that is common and widespread in gardens, thriving in disturbed ground; it frequently appears in soil cultivated for vegetables. A close look at the stems reveals lines of hairs on alternate sides between the leaf nodes. Common Chickweed flowers are white, 5-petalled and 5–10mm across; it has a long flowering period and, although commonest in summer, flowers can be found in any month. The leaves are oval, fresh green and in opposite pairs.

LESSER STITCHWORT

Stellaria graminea HEIGHT TO 50CM

Attractive plant of grassy woodland rides, meadows and hedgerows; it is common and widespread but does best on acid soils and sometimes persists around the margins of rural gardens. Unlike Greater Stitchwort (see above) it has smooth-edged stems. Lesser Stitchwort flowers are white and 5–15mm across, with 5 deeply divided petals; they appear May–August, often in profusion. Its leaves are long, narrow, smooth-edged and grass-like; they are easy to overlook.

COMMON MOUSE-EAR

Cerastium fontanum HEIGHT TO 30CM

Hairy plant that produces both flowering and non-flowering shoots. It is common and widespread throughout Britain, favouring disturbed ground, arable fields and track margins in particular; it often turns up in cultivated ground in the garden, sometimes growing in profusion. Common Mouse-ear flowers are white, 5–7mm across with 5 deeply notched petals; they appear April–October. Its leaves are grey-green and opposite. The seeds are eaten by finches.

RED CAMPION

Silene dioica HEIGHT TO 1M

Familiar and attractive, hairy plant that is common and widespread in hedgerows, grassy banks and wayside places generally; it sometimes appears in parts of the garden that have been set aside for native wildflowers. Red Campion flowers are reddish pink and 20-30mm across; male flowers are smaller than females and borne on separate-sex plants. The plant has an extended flowering period, blooms being found March-October in most parts of Britain. Its leaves are hairy and borne in opposite pairs.

WINTER ACONITE

Eranthis hyemalis HEIGHT TO 10CM

Attractive member of the buttercup family that is introduced to Britain, planted for its winter colour; it is now widely naturalised in woodlands and shady places and sometimes forms carpets on woodland floors. Winter Aconite flowers are 12-15mm across, with 6 yellow sepals; they are borne on upright stems, standing proud of the leaves, and appear January-April. Its leaves are spreading and each is divided into 3 lobes; there are 3 leaves per stem.

CREEPING BUTTERCUP

Ranunculus repens
HEIGHT TO 50CM

Widespread and common plant of damp, grassy places in the countryside; it is a familiar and often unwelcome perennial of garden lawns and its will to survive often dismays gardeners. Creeping Buttercup flowers are 2-3cm across with 5 yellow petals and upright sepals; they are borne on furrowed stalks and appear May–August. The fruits are borne in rounded heads. The leaves are hairy and divided into 3 lobes (middle lobe stalked).

LESSER SPEARWORT

Ranunculus flammula
HEIGHT TO 50CM

Variably upright or creeping plant that grows in damp ground, often at the margins of water; it often roots where leaf nodes touch the ground. It will grow around the edges of well-established garden ponds with muddy margins, or in bog gardens. Lesser Spearwort flowers are 5-15mm across with 5 golden yellow petals; they are borne on slender stalks and appear June-October. The leaves are variable: basal leaves are oval, stem leaves are narrow.

WHITE CAMPION

Silene latifolia HEIGHT TO 1M

Attractive wayside plant that is hairy and branched; it sometimes hybridises with Red Campion (see above). It is common and widespread, being found in a range of grassy and disturbed habitats, and sometimes turns up in parts of the garden devoted to native wildflowers. White Campion flowers are white, 5-petalled and 25-30mm across; male flowers are smaller than females and borne on separate sex plants, appearing May-October. Its leaves are oval and opposite.

MEADOW BUTTERCUP

Ranunculus acris HEIGHT TO 1M

Widespread and familiar, downy plant that is often abundant in damp, grassy habitats; it sometimes thrives in grassy areas of the garden set aside for native wildflowers. Meadow Buttercup flowers are 18-25mm across, with 5 shiny, yellow petals and upright sepals; they are borne on smooth stalks and appear April-October. Its fruits are hook-tipped and borne in rounded heads, and the leaves are rounded in outline but divided into 3-7 lobes; the upper leaves are unstalked.

BULBOUS BUTTERCUP

Ranunculus bulbosus
HEIGHT TO 40CM

Widespread and locally common buttercup of dry, free-draining soils and is particularly common in chalk grassland. It sometimes occurs in gardens, but only where soil conditions suit its need. Bulbous Buttercup flowers are 2-3cm across with 5 bright yellow petals and reflexed sepals; they are borne on furrowed stalks and appear March–July. The fruits are smooth and the leaves have 3 lobes (all stalked).

LESSER CELANDINE

Ranunculus ficaria HEIGHT TO 25CM

Patch-forming perennial that is common and widespread in woodland and hedgerows throughout Britain. It often spreads into the margins of rural gardens and provides ground cover and an excellent splash of colour in early spring. Lesser Celandine flowers are 2-3cm across with 8-12 shiny yellow petals and 3 sepals; they are borne on long stalks and appear March-May, opening only in bright sunshine. The leaves are heart-shaped, glossy and dark green.

WILDFLOWERS

Cabbage family members are staple components of the vegetable plot while roses are ever popular in flower borders. Both families have numerous wildflower representatives that often stray into the garden too.

HEDGE MUSTARD
Sisymbrium officinale HEIGHT TO 90CM

Tough, upright plant that is common and widespread throughout Britain, growing in disturbed soil and on waste ground and roadside verges; it sometimes appears in gardens where the ground has been disturbed recently. Hedge Mustard flowers are 3mm across with 4 yellow petals; they are borne in terminal clusters and appear May–October. The fruits are cylindrical, 1–2cm long and pressed close to the stem. Leaves are variable: lower ones deeply divided, upper ones narrow.

WINTER-CRESS
Barbarea vulgaris HEIGHT TO 80CM

Upright, hairless plant with edible, albeit peppery, leaves. It is common and widespread in the south and does best on damp ground; it often appears in gardens within its range. Winter-cress flowers are 7–9mm across with 4 yellow petals; they are borne in terminal heads May–August. The fruits are long, narrow and 4-sided pods and the leaves are dark green and shiny; the lower ones are divided with a large, oval end lobe, while the upper stem leaves are undivided.

HAIRY BITTER-CRESS
Cardamine hirsuta HEIGHT TO 30CM

Upright annual plant with hairless stems. It is common and widespread and grows best in damp, disturbed ground. Hairy Bitter-cress flowers are 2–3mm across and terminal; the extended flowering season means that flowers are seen February–November. Note that petals are absent in some flowers. The fruits are curved, up to 2.5cm long and overtop the flowers. Its leaves are deeply divided with rounded lobes; they are produced mainly as a basal rosette with a further 1–4 leaves up the stem.

SHEPHERD'S-PURSE
Capsella bursa-pastoris HEIGHT TO 35CM

Common plant with distinctive fruits. In the countryside it grows in arable fields, beside tracks and on waste ground; in the garden it is common in the vegetable patch. Shepherd's-purse flowers are 2–3mm across with 4 white petals; they are borne in terminal clusters and can be found throughout the year. Its fruits are green, triangular and notched – they are fancifully purse-like. The leaves vary from lobed to entire; upper ones are usually toothed and clasp the stem.

CUCKOOFLOWER
Cardamine pratensis HEIGHT TO 50CM

Attractive, variable plant that is also known as Lady's-smock. It is locally common and widespread, growing best in damp, grassy places; it sometimes persists in damp patches of lawn where the mowing regime is sympathetic. Cuckooflower flowers are 12–20mm across with 4 pale lilac or white flowers; they appear April–June. Its fruits are elongated and beaked and the leaves are seen mainly in a basal rosette of deeply divided leaves with rounded lobes; narrow stem leaves are also present.

COMMON WHITLOWGRASS
Erophila verna HEIGHT TO 20CM

Variable, hairy annual plant that is common and widespread in dry, bare places. In the garden, it often appears on gravel drives or on well-drained bare soil, and can be abundant. Common Whitlowgrass flowers are 3–6mm across and comprise 4 deeply notched whitish petals; they appear March–May. The fruits are elliptical pods, borne on long stalks. Its leaves are narrow and toothed; they form a basal rosette from the centre of which the flowering stalk arises.

Hedge Mustard   **Winter-cress**   **Cuckooflower**

THALE CRESS

Arabidopsis thaliana
HEIGHT TO 50CM

Distinctive plant that is tough and resilient despite its extremely slender appearance. It is locally common and grows best on dry, sandy soils; in the garden it sometimes appears on gravel paths, in cracks in paving and in flowerpots. Thale Cress flowers are 3mm across with 4 white petals; they are borne in terminal clusters and appear March–October. The fruits are cylindrical and 20mm long and the leaves are broadly toothed, oval and form a basal rosette.

AGRIMONY

Agrimonia eupatoria
HEIGHT TO 50CM

Upright plant of grassy places, hedgerows and roadside verges; it is common and widespread and sometimes turns up on disturbed soil in gardens. Agrimony flowers are 5–8mm across with 5 yellow petals; they are borne in upright, swaying spikes and appear June–August. The fruits are bur-like and covered in spines; they catch on clothes, and on animal fur, to assist dispersal. Its leaves comprise 3–6 pairs of oval, toothed leaflets with smaller leaflets between the larger ones.

BRAMBLE

Rubus fruticosus HEIGHT TO 3M

Familiar scrambling shrub that is common and widespread in hedgerows and scrub, and in gardens too. Its appearance is rather variable but all forms have arching stems armed with variably shaped prickles; stems root when they touch the ground. Bramble flowers are 2–3cm across and white or pink; they appear May–August. The fruits are familiar blackberries and the leaves have 3–5 toothed leaflets. Well worth tolerating for its fruits and for the nesting opportunities its tangled stems provide.

SILVERWEED

Potentilla anserina CREEPING

Low-growing plant that has long, creeping stems that are sometimes tinged reddish. It is common and widespread, favouring damp, grassy places and bare ground and it often appears in gardens, colonising bare ground such as vegetable patches. Silverweed flowers are 15–20mm across with 5 yellow petals; they appear May–August. Its leaves are divided into up to 12 pairs of leaflets (with tiny ones between them) that are covered in the silvery, silky hairs that give the plant its name.

ABOVE:
flowers
LEFT:
fruits

GARLIC MUSTARD

Alliaria petiolata HEIGHT TO 1M

Familiar, robust wayside plant that is locally abundant in hedgerows and on roadside verges. It often strays into rural gardens and is worth encouraging because its leaves are the foodplant for the larvae of Orange-tip butterflies. Garlic Mustard flowers are 6mm across with 4 white petals; they appear April–June. Its fruits are cylindrical, ribbed and 4–5cm long and the leaves are heart-shaped, toothed and borne up stem; they smell strongly of garlic when crushed.

DOG-ROSE

Rosa canina HEIGHT TO 3M

Scrambling, variable shrub with long, arching stems that bear curved thorns. It is common and widespread in hedgerows and scrub, and sometimes occurs naturally, or is planted, in garden hedgerows to add to their native appeal. Dog-rose flowers are 3–5cm across and fragrant with 5 pale pink petals; they are borne in clusters of up to 4 flowers and appear June–July. Its fruits are red, egg-shaped hips that typically shed their sepals before they ripen. The leaves comprise 5–7 hairless leaflets.

TORMENTIL

Potentilla erecta HEIGHT TO 30CM

Creeping, downy plant that is widespread and locally common in grassy places, especially on well-drained soils of heaths and moors; it sometimes occurs on lawns in rural gardens with suitable soils. Tormentil flowers are 7–11mm across with 4 yellow petals (petals are easily dislodged); the flowers are borne on slender stalks and appear May–September. Its leaves are unstalked and trifoliate, but appear 5-lobed because of the two large, leaflet-like stipules at the base.

WOOD AVENS

Geum urbanum
HEIGHT TO 50CM

Hairy plant that is common in hedgerows and woodlands. It is also a familiar garden plant, growing on gravel tracks and in borders. Wood Avens flowers are 8–15mm across and comprise 5 yellow petals; they are upright in bud but drooping when fully open May–August. Its fruits are bur-like, with red, hooked spines that catch on clothes and animal fur. Basal leaves have 3–6 pairs of side leaflets and a large terminal one; stem leaves are 3-lobed.

ABOVE: **flowers**
RIGHT: **fruits**

Clovers, trefoils and vetchs are members of the pea and bean family as can be confirmed by looking at the pods in which their seeds are produced. Crane's-bills and members of the spurge family have both native and garden representatives and the distinction between the two is often blurred in horticultural terms: many native species are used widely to provide colour in herbaceous borders.

TUFTED VETCH

Vicia cracca HEIGHT TO 2M

Slightly downy, scrambling plant of grassy places, hedgerows and scrub; it sometimes grows around the margins of rural gardens and makes an interesting addition to any herbaceous border. Tufted Vetch flowers are 8-12mm long and bluish purple; they are borne in one-sided spikes up to 8cm tall, appear June–August and are popular with pollinating insects. Its fruits are hairless pods and the leaves comprise up to 12 pairs of narrow leaflets and end in a branched tendril.

RIBBED MELILOT

Melilotus officinalis HEIGHT TO 1.5M

Attractive, upright and hairless plant that is probably introduced to Britain but now grows in a wide range of grassy places, especially on waste ground and roadside verges. It sometimes appears in gardens, especially on patches of disturbed soil. Ribbed Melilot flowers are bright yellow and borne in spikes up to 7cm long; they appear June–September and are popular with pollinating insects. Its fruits are brown, wrinkled pods and the leaves comprise 3 oblong leaflets.

COMMON BIRD'S-FOOT TREFOIL

Lotus corniculatus HEIGHT TO 10CM

Sprawling, solid-stemmed and usually hairless plant that is common in grassland. It sometimes grows in gardens, especially in grassy areas set aside for native wildflowers. Common Bird's-foot Trefoil flowers are red in bud but yellow and 15mm long when open; they are borne in heads on stalks to 8cm long and appear May–September. Its fruits are slender pods that splay like a bird's foot when ripe. The leaves have 5 leaflets but usually appear trefoil.

HOP TREFOIL

Trifolium campestre
HEIGHT TO 25CM

Low-growing, hairy plant that is common and widespread in dry grassland on free-draining soils; if soil conditions and mowing regime suit it, it will grow in garden lawns. Hop Trefoil flowers are 4-5mm long and yellow; they are borne in compact, rounded heads, 15mm across and appear May–October. Its fruits are pods, cloaked by brown dead flowers; the whole resembles miniature hop-like heads. The leaves are trefoil and the terminal leaflet has the longest stalk.

flowers   **fruit**

BLACK MEDICK

Medicago lupulina HEIGHT TO 20CM

Downy, usually low-growing annual that is common in short grassland and on waste ground; it grows quite happily in garden lawns, and tolerates mowing. Black Medick flowers are small and yellow; they are borne in dense, spherical heads (8-9mm across) that comprise 10-50 flowers, and they appear April–October. Its fruits are spirally coiled and spineless, green at first but ripening black. The leaves are trefoil, each leaflet bearing a point at the centre of its apex.

RED CLOVER

Trifolium pratense HEIGHT TO 40CM

Familiar downy plant that is common and widespread in grassy places on a wide range of soil types. It is sometimes found in the garden, especially in grassy areas devoted to native wildflowers or lawns not subjected to harsh mowing regimes. Red Clover flowers are pinkish red and are borne in dense, unstalked heads, 2-3cm across; they appear May–October and are visited by pollinating insects. Its leaves are trefoil and the oval leaflets each bear a white, crescent-shaped mark.

WHITE CLOVER

Trifolium repens HEIGHT TO 40CM

Creeping, hairless plant that roots at the nodes. It is common and widespread in grassy areas on a wide range of soil types and sometimes grows in gardens, in lawns or grassy patches. White Clover flowers are creamy white at first, browning with age; they are borne in long-stalked rounded heads, 2cm across and appear May–October. The leaves are trefoil and often each of the 3 rounded leaflets has a central white mark and translucent lateral veins.

CUT-LEAVED CRANE'S-BILL

Geranium dissectum HEIGHT TO 45CM

Straggly, hairy plant that is common and widespread except in the north of Britain, growing on disturbed ground and in cultivated soil in the countryside. In the garden, it sometimes turns up in the tilled vegetable patch or on disturbed ground. Cut-Leaved Crane's-bill flowers are 8–10mm across with pink, notched petals; they are borne on short stalks and appear May–September. Its fruits are downy; the leaves are deeply cut.

SUN SPURGE

Euphorbia helioscopia HEIGHT TO 50CM

Upright, hairless and yellowish-green plant that is common and widespread on disturbed ground and in cultivated soils in farmland areas. In the garden, tilled vegetable patches and the margins of paths provide it with ideal growing conditions. Sun Spurge flowers lack sepals and petals and are yellow with green lobes; they are borne in flat-topped umbel-like clusters with 5 leaf-like basal bracts and appear May–November. Its leaves are spoon-shaped and toothed.

ANNUAL MERCURY

Mercurialis annua HEIGHT TO 50CM

Hairless, branched and bushy annual plant that is locally common in southern Britain, growing in cultivated soils and on waste ground; it does especially well near the sea. It is frequent in gardens within its range. Its rather nondescript appearance means that it is often overlooked. Annual Mercury flowers are yellowish green and borne in spikes on separate sex plants; they appear July–October. Its fruits are bristly and the leaves are narrowly ovate, shiny and toothed.

HERB-ROBERT

Geranium robertianum HEIGHT TO 30CM

Straggling, hairy plant that is common in shady places in hedgerows, verges and woodlands. It grows in similar, suitable shady places in the garden and its dainty flowers provide interesting splashes of subdued colour from spring to autumn. Herb-Robert flowers are 12–15mm across with pink petals and orange pollen; they are borne in loose clusters and appear April–October. The fruits have a long 'beak' and the leaves are hairy and deeply divided; they are often tinged red.

DOVE'S-FOOT CRANE'S-BILL

Geranium molle HEIGHT TO 20CM

Spreading and branched, extremely hairy plant that is common and widespread, especially in the south. It favours dry, grassy places including roadside verges and grazed and lightly trampled meadows; it can sometimes be found on garden lawns if these are managed sympathetically. Dove's-foot Crane's-bill flowers are 5–10mm across with notched pink petals; they are borne in pairs and appear April–August. Its fruits are hairless and the leaves are hairy and rounded, with the margins cut into 5–7 lobes.

PETTY SPURGE

Euphorbia peplus HEIGHT TO 30CM

Upright, hairless plant that often branches from the base. It is common and widespread throughout and favours arable land and cultivated fields, where it can be abundant. In the garden, it grows in borders, vegetable patches and in soil-filled flowerpots. Petty Spurge flowers are greenish with oval bracts (sepals and petals are absent); they are borne in flattish umbel-like clusters and appear April–October. Its fruits are smooth and the leaves are oval, blunt-tipped and stalked.

DOG'S MERCURY

Mercurialis perennis HEIGHT TO 35CM

Hairy, creeping perennial with a foetid smell. In the countryside at large, it grows in woodlands and forms carpets where conditions suit it and it is not disturbed. Its presence around the margins of rural gardens often indicates that woodland was cleared to create the garden. Dog's Mercury flowers are yellowish and rather tiny; they are borne in open spikes on separate sex plants and appear February–April. Its fruits are hairy and the leaves are oval, shiny and toothed.

Tolerate or encourage wildflowers in the garden and you will have a colourful display from early spring to late autumn. You can also relax in the knowledge other forms of native wildlife will benefit from their presence too. However, a few need to be kept in check and a small minority can be positively invasive.

SWEET VIOLET

Viola odorata HEIGHT TO 15CM

Attractive, fragrant perennial herb that is common and widespread in England and Wales; it favours woods and hedgerows, mostly on calcareous soils. It is widely planted in gardens, thriving if soil conditions suit it, and does best in partial shade. Sweet Violet flowers are 15mm across and violet or white, with blunt sepals; they appear February–May. Its fruits are egg-shaped and the leaves are long-stalked and rounded in spring; larger and heart-shaped leaves appear in autumn.

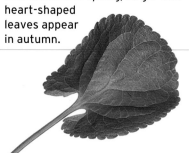

COMMON DOG-VIOLET

Viola riviniana HEIGHT TO 12CM

Familiar, patch-forming perennial herb that is common and widespread throughout much of Britain, growing in woodland and shaded grassland, and on roadside verges. It also grows in rural gardens, and does best in partial shade. Common Dog-violet flowers are 15–25mm across and bluish violet with a blunt, pale spur that is notched at the tip, and pointed sepals; they appear March–May. Fruits are egg-shaped; leaves are long-stalked and heart-shaped.

HOARY WILLOWHERB

Epilobium parviflorum
HEIGHT TO 75CM

Upright, perennial plant that is similar to Rosebay Willowherb but smaller and downy, with spirally arranged, not alternate, stem leaves. It is common and widespread, except in the north, and favours damp, disturbed ground; it grows happily in damp herbaceous borders in the garden. Hoary Willowherb flowers are 12mm across with pale pink, notched petals and a 4-lobed stigma; they appear July–September. Its fruits are pods that contain cottony seeds and the leaves are broadly oval.

ROSEBAY WILLOWHERB

Chamerion angustifolium
HEIGHT TO 1.5M

Showy perennial that is common and widespread throughout Britain, growing on waste ground, cleared woodland and riverbanks, on a wide range of soil types. It sometimes grows in disturbed ground in the garden. Rosebay Willowherb flowers are 2–3cm across with pinkish-purple petals; they are borne in tall spikes and appear July–September. Its fruits are pods that contain cottony seeds and the leaves are narrow and arranged spirally up the stems; they are eaten by Elephant Hawkmoth larvae.

Rosebay Willowherb

Hoary Willowherb

BROAD-LEAVED WILLOWHERB

Epilobium montanum
HEIGHT TO 80CM

Upright perennial plant that is similar to Hoary Willowherb but almost hairless. It is common and widespread, and grows in woods and hedgerows. It also occurs in the garden, turning up in herbaceous borders and vegetable plots. Hoary Willowherb flowers are 6–10mm across (they droop in bud) with pale pink, notched petals and a 4-lobed stigma; they appear June–August. Its fruits are pods with cottony seeds; the leaves are oval.

AMERICAN WILLOWHERB

Epilobium ciliatum
HEIGHT TO 50CM

Upright perennial plant, introduced to Britain but now widespread and common. It grows in damp, shady places, and often turns up in the garden, sometimes appearing in the vegetable patch or in potted compost; its spread is probably aided by the sale of garden centre plants. American Willowerb flowers are 8–10mm across with pink, notched petals and a club-shaped stigma; they appear July–September. Its leaves are narrow-oval and short-stalked.

COW PARSLEY
Anthriscus sylvestris
HEIGHT TO 1M

Downy, herbaceous plant with hollow, unspotted stems. It is common and widespread in meadows and woodland margins, and particularly noticeable when in flower on roadside verges; it often appears in the garden too. Cow Parsley flowers are white and borne in umbels up to 6cm across; they appear April–June and are extremely popular with insects. Its fruits are elongate and ridged and the leaves are deeply divided 2 to 3 times, only slightly hairy and fresh green.

GROUND-ELDER
Aegopodium podagraria
HEIGHT TO 1M

Patch-forming plant, originally introduced as a pot and medicinal herb and now rampantly invasive in many gardens. The bane of many gardeners' lives, it thrives on disturbance (broken stems grow into new plants) and so cultivating the soil can positively benefit it. Ground-elder flowers are white and borne in umbels, 2–6cm across with 10–20 rays; they appear May–July and are popular with insects. Its fruits are egg-shaped and the leaves are roughly triangular.

PRIMROSE
Primula vulgaris HEIGHT TO 20CM

Familiar and attractive herbaceous plant that is common and widespread in hedgerows, woodlands and shady meadows. But it also does well in gardens, where it is often planted and encouraged. Primrose flowers are 2–3cm across, 5-lobed and pale yellow, usually with deep yellow centres; they are solitary and borne on hairy stalks that arise from centre of the leaf rosette, appearing February–May. The leaves are oval, tapering and crinkly, and up to 12cm long; they form a basal rosette.

CREEPING-JENNY
Lysimachia nummularia
CREEPING

Low-growing, hairless plant that is locally common in England but scarce or absent elsewhere. It likes to grow in damp, grassy ground and is often found in shady spots in the garden, especially areas where water collects or where drainage is poor. Creeping-Jenny flowers are 15–25mm across, yellow and bell-shaped with 5 pointed lobes; they are borne on stalks arising from leaf axils and appear June–August. Its rounded or heart-shaped leaves appear in opposite pairs.

HOGWEED
Heracleum sphondylium
HEIGHT TO 2M

Robust, roughly hairy plant with hollow, ridged stems. It is common and widespread in meadows, open woodlands and roadside verges, but is also present in many rural gardens. Hogweed flowers are off-white, with unequal petals, and are borne in umbels up to 20cm across, with 40 or so rays; they appear May–August and are extremely popular with insects. Its fruits are elliptical and flattened and the leaves are up to 60cm long, broad, hairy and divided into oval lobes.

FOOL'S-PARSLEY
Aethusa cynapium
HEIGHT TO 50CM

Delicate, hairless and unpleasant tasting annual plant. It is commonest in the south and grows in disturbed soils of arable fields and gardens; it does particularly well in the regularly tilled ground of vegetable patches. Fool's-parsley flowers are white and borne in umbels, 2–3cm across; the secondary umbels have a 'beard' of long upper bracts and the flowers appear June–August. Its fruits are egg-shaped; the leaves are twice divided, flat and triangular.

COWSLIP
Primula veris HEIGHT TO 25CM

Elegant, downy plant that is locally common and widespread in dry grassland, often on calcareous soils, typically where the habitat is 'unimproved' (has not been ploughed or sprayed). It also does well in the garden, often spreading from herbaceous borders into neighbouring lawns. Cowslip flowers are 8–15mm across, fragrant, bell-shaped, stalked and orange-yellow; they are borne in rather 1-sided umbels of 10–30 flowers and appear April–May. Its tapering, wrinkled leaves form a basal rosette.

YELLOW PIMPERNEL
Lysimachia nemorum CREEPING

Evergreen, hairless plant that is superficially similar to Creeping-Jenny but more delicate. It is common and widespread, growing in damp, shady places, especially in woodlands. It is also found in those gardens that provide it with suitable habitats. Yellow Pimpernel flowers are 10–15mm across, yellow and star-shaped with 5 lobes; they are borne on slender stalks arising from leaf axils and appear May–August. Its leaves are oval or heart-shaped, and are borne in opposite pairs.

A number of native wildflower species grow in the countryside in the disturbed soils, around the margins of fields and in shady hedgerows. Little wonder then that they also turn up closer to home when gardeners create similar habitats in the vegetable patch, flower border or shady hedge.

SCARLET PIMPERNEL

Anagallis arvensis CREEPING

Low-growing, hairless annual that is common and widespread in cultivated and disturbed ground. It does well in the garden and is a frequent discovery in even the best tended of vegetable patches. Scarlet Pimpernel's charming flowers are 10-15mm across with 5 scarlet or pinkish orange petals that are fringed with hairs; the flowers open wide only in sunshine (remaining tightly closed on rainy or dull days) and appear June-August. Its leaves are oval and usually in pairs.

ABOVE: **Scarlet Pimpernel**
BELOW: **Hedge Bindweed**

FIELD BINDWEED

Convolvulus arvensis
HEIGHT TO 3M

Familiar perennial that twines around other plants to assist its progress through vegetation. It is widespread and common almost everywhere except in northern Britain and favours disturbed and arable land; it is a persistent garden weed, found both in herbaceous borders and vegetable patches. Field Bindweed's attractive flowers are 15-30mm across, funnel-shaped and either white or pink with broad, radiating white stripes; the flowers appear June-September. Its leaves are easily recognised, being arrow-shaped, 2-5cm long and long-stalked.

HEDGE BINDWEED

Calystegia sepium HEIGHT TO 5M

Unable to support its own weight, Hedge Bindweed achieves an impressive stature in its favoured hedgerow habitat by twining around other plants to assist its sky bound progress. It is common and widespread in much of lowland Britain and often does well in garden hedges. Its flowers are 3-4cm across, white and funnel-shaped, and they appear June-September. There are 2 green bracts at the base of the flowers and the leaves are arrow-shaped and up to 12cm long.

CLEAVERS

Galium aparine HEIGHT TO 1.5M

Sprawling wayside plant that is common in hedgerows, and frequent in the garden; if left untended, it can quickly smother less vigorous plants. Its leaves and square stems have backward-pointing bristles on the edges that aid its scrambling progress through vegetation. Cleavers flowers are 2mm across and greenish white, with 4 petals; they are borne in clusters arising from leaf axils and appear May-September. Its fruits are nutlets with hooked bristles; they catch in clothes and animal fur.

ABOVE: **Cleavers fruit**
BELOW: **Early Forget-me-not**

COMMON COMFREY

Symphytum officinale
HEIGHT TO 75CM

Rough, hairy wayside plant that favours damp ground and grows beside rivers and ditches. It is common in central and southern England and is also grown in much of Britain in gardens, sometimes coming to dominate borders if left untended. Common Comfrey has strikingly winged stems and its flowers, borne in curved clusters, are 12-18mm long and white, pink or purple; they appear May-June. Its oval, hairy leaves have a stalk that runs down the main stem.

EARLY FORGET-ME-NOT

Myosotis ramosissima
HEIGHT TO 10CM

Downy annual plant that grows in arable fields and bare grassy places; it sometimes appears in the garden, typically growing in areas of short or worn grass. It is common and widespread except in the north of Britain. Early Forget-me-not's miniature flowers are 2-3mm across, 5-lobed and sky blue; they are borne in clusters and appear April-October; the petals are shorter than the sepal tube. Its leaves are oval, the basal ones forming a rosette.

SELFHEAL

Prunella vulgaris HEIGHT TO 20CM

Downy plant with leafy rooting runners and upright flowering stems. It is commonest in the south and it grows in meadows and on verges. Where conditions suit it (it likes calcareous or neutral soils), it is found in gardens, adding an uninvited splash of native colour to herbaceous borders. Its flowers are 10–15mm long and bluish violet; they are borne in dense, cylindrical heads with purplish bracts and calyx teeth, and appear April–June. The paired leaves are oval.

WHITE DEAD-NETTLE

Lamium album
HEIGHT TO 40CM

Downy, slightly aromatic and patch-forming perennial plant that grows in meadows and hedgerows, and on roadside verges; it is common and widespread in much of Britain. White Dead-nettle flowers are 25–30mm long and white, with a hairy upper lip and toothed lower lip; they are borne in whorls, appear March–November, and are popular with insects. Its leaves are heart-shaped to oval and have toothed margins; they resemble those of Common Nettle but they lack stinging hairs.

HENBIT DEAD-NETTLE

Lamium amplexicaule
HEIGHT TO 20CM

Trailing, straggly annual plant that is fairly common in cultivated soil and disturbed ground, typically in dry locations on free-draining soils; where conditions meet its requirements, it also occurs in the garden. Henbit Dead-nettle flowers are 15–20mm long and pinkish purple with a hairy lip and long corolla tube; they are borne in widely spaced whorls and appear March–November; only a few flowers in a given whorl open at any one time. The leaves are rounded and blunt-toothed.

BITTERSWEET

Solanum dulcamara
HEIGHT TO 1.5M

Scrambling member of the nightshade and potato family that has a woody base. It is commonly found in hedgerows and scrub and it often grows around the margins of rural gardens, sometimes on manure-enriched soil. Bittersweet flowers are 10–15mm across with 5 purple, petal-like corolla lobes and projecting yellow anthers; they are borne in hanging clusters on purple stems and appear May–September. Its fruits are poisonous red berries; the leaves are oval and pointed.

GROUND-IVY

Glechoma hederacea
HEIGHT TO 15CM

Softly hairy, aromatic perennial plant with creeping, rooting runners and upright flowering stems. It is common and widespread in much of Britain, growing in woods, hedgerows and meadows; where conditions suit it, it carpets the ground and it often grows in rural gardens too, appearing in shady spots and in herbaceous borders. Ground-ivy flowers are 15–20mm long and bluish violet; they are borne in open whorls arising from leaf axils. The leaves are kidney-shaped to rounded, toothed and long-stalked.

RED DEAD-NETTLE

Lamium purpureum
HEIGHT TO 30CM

Downy, pungently aromatic plant whose leaves and stems are sometimes tinged purplish. It is common and widespread in disturbed ground and in cultivated soils; unsurprisingly, it is often seen in the garden, favouring tilled soils of the vegetable patch. Red Dead-nettle flowers are 12–18mm long and purplish pink, with a hooded upper lip and lower lip toothed at the base; they are borne in whorls on upright stems and appear March–October. Its leaves are heart-shaped.

SPEAR MINT

Mentha spicata HEIGHT TO 75CM

Familiar culinary herb and the most popular cultivated mint to be found in the garden. It favours damp ground and where conditions suit it, the plant often spreads rampantly through borders; occasionally it even becomes naturalised in the countryside at large. Spear Mint has pinkish-lilac flowers that are borne in whorled terminal spikes and appear July–October; they are popular with insects. Its leaves are narrowly oval, toothed and almost unstalked, and they smell strongly of spearmint when bruised.

COMMON FIGWORT

Scrophularia nodosa
HEIGHT TO 70CM

Robust plant whose tall stems are square in cross section. It is common and widespread in much of Britain, growing in damp and shady spots in woodland and hedgerows; it turns up in similar habitats in the garden. Common Figwort flowers are small but they are beautiful when viewed in close-up, being greenish with a maroon upper lip; its fruits bear a fanciful resemblance to miniature figs. It has oval, pointed leaves whose margins have sharp teeth.

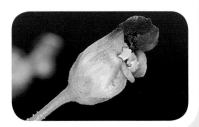

Many wildflowers have a tenacious will to live, demonstrated by the fact that some thrive in disturbed ground, some even tolerating a degree of trampling. Where similar conditions are created in the garden – in lawns, paths and borders – these same plants appear, welcome or otherwise.

GERMANDER SPEEDWELL

Veronica chamaedrys
HEIGHT TO 20CM

Delicate perennial with creeping, rooting stems and upright flowering stems. It is common and widespread, favouring a wide range of grassy places; in the garden it sometimes grows in lawns where the mowing regime is not too strict. Germander Speedwell flowers are 10-12mm across and blue with a white centre; they are borne on slender stalks in open, terminal spikes and appear April-June. Its fruits are heart-shaped capsules and the leaves are oval, toothed, hairy and short-stalked.

SLENDER SPEEDWELL

Veronica filiformis PROSTRATE

Mat-forming, downy perennial with creeping stems. Although introduced to Britain, it is now locally common; it grows in the countryside at large but often does best on garden lawns, providing sizeable patches of colour in the spring, before regular mowing starts. Slender Speedwell flowers are 8-10mm across and bluish with a white lip; they are borne on long, slender stalks arising from leaf axils and appear April-July. Its leaves are 5-10mm across, rounded to kidney-shaped, blunt-toothed and short-stalked.

FOXGLOVE

Digitalis purpurea HEIGHT TO 1.5M

Much loved plant that is common in woods and moors, especially on acid soils, but equally at home in gardens; here it makes a colourful contribution to any herbaceous border. Foxglove flowers are 4-5cm long, pinkish purple (sometimes white) with darker spots in the throat; they are borne in tall spikes, appear June-September, and are visited by bumblebees. The leaves are 20-30cm long, downy and oval; they form a rosette in the 1st year from which the flowering spike appears in the 2nd.

GREATER PLANTAIN

Plantago major HEIGHT TO 20CM

Persistent and usually hairless plant that is extremely common, growing on disturbed and trampled ground, tracks and arable fields; in the garden, it is a frequent, if unwelcome, component of many lawns. Greater Plantain flowers are 3mm across, pale yellow with anthers that are purple at first, turning yellow later; they are borne on slender spikes, 10-15mm long, and appear June-October. Its leaves are broad, oval, up to 25cm long, with 3-9 veins and a narrow stalk.

LEFT: **Greater Plantain**

RIBWORT PLANTAIN

Plantago lanceolata
HEIGHT TO 15CM

Persistent perennial plant that is common and widespread in much of Britain, growing on disturbed grassland and tracks. It often grows in the garden, favouring tracks and paths and tolerating (indeed benefiting from) trampling and mowing. Ribwort Plantain flowers are 4mm across and brownish with white stamens; they are borne in compact heads, 2cm long, on furrowed stalks up to 40cm long, and appear April-October. Its leaves are narrow, with 3-5 veins.

WILD TEASEL

Dipsacus fullonum HEIGHT TO 2M

Biennial plant with prickly stems. Thrives on damp, disturbed ground in the countryside. Often planted in gardens for its spiny, architectural seed heads; these attract Goldfinches, which feed on the seeds. Teasel flowers are pinkish purple; they are borne in egg-shaped heads, 6-8cm long, adorned with spiny bracts, and appear July-August. Its leaves are spine-coated; they form rosettes in the 1st year while in the 2nd year they are seen as opposite stem leaves fused at the base and collecting water.

DAISY

Bellis perennis HEIGHT TO 10CM

Familiar, downy perennial plant. Although it grows in short grassland of all types, the species does best in lawns, tolerating mowing even if this means the flowers seldom get the chance to set seed. Daisy flowers are borne in solitary heads, 15–25mm across; they are carried on slender stems, comprise yellow inner florets and white (often faintly crimson-tipped) outer florets, and they appear March–October. Its leaves are spoon-shaped and form prostrate rosettes.

PINEAPPLE MAYWEED

Matricaria discoidea HEIGHT TO 12CM

Bright green, hairless perennial that smells distinctly of pineapple when the leaves are crushed. It is common and widespread in the countryside, being a plant of disturbed ground, tracks and paths, and tolerating a degree of trampling. In the garden, it sometimes turns up in similar locations. Pineapple Mayweed flowers comprise yellowish-green florets that are borne in rounded to conical, hollow heads, 8–12mm long; they appear May–November. Its leaves are feathery.

YARROW

Achillea millefolium HEIGHT TO 50CM

Downy, aromatic perennial plant that has creeping stems and upright flowering stalks. It is common and widespread in grassy places such as meadows, roadside verges and waste ground, and it sometimes appears on disturbed grassy ground in the garden. Yarrow flowers are borne in flat-topped clusters of heads, each 4–6mm across, with yellowish central florets and pinkish-white outer ones; they appear June–November. Leaves are feathery.

TANSY

Tanacetum vulgare HEIGHT TO 75CM

Upright, aromatic perennial plant that is common and widespread in hedgerows and on roadside verges; it is also popular in the garden and its flower heads are valued for the strikingly colourful addition they make to the herbaceous border and because of their popularity with insects. Tansy flowers are borne in flat-topped, umbel-like clusters of up to 70 golden yellow, button-like heads, 7–12mm across; they appear July–October. Its leaves are deeply divided.

SCENTLESS MAYWEED

Tripleurospermum inodorum HEIGHT TO 75CM

Scentless, hairless and straggly perennial plant. Common in cultivated and disturbed ground, mainly in farmland, but occasionally appears in quantity after garden soil has been tilled, if the seedbank has been disturbed. Scentless Mayweed flowers are borne in clusters of stalked heads, each 20–40mm across and resembling large Daisy flower heads; they comprise yellow inner florets and white outer florets; they appear April–October. The leaves are feathery.

COMMON FLEABANE

Pulicaria dysenterica HEIGHT TO 50CM

Creeping perennial plant with upright, woolly flowering stems. It is common and widespread, growing in damp, grassy places, usually on heavy soils; if conditions suit it, it does well in the garden and its flowers are extremely popular with insects. Common Fleabane flowers are borne in clusters of heads, 15–30mm across, with spreading yellow outer florets and deeper yellow, central florets; they appear July–September. Its leaves are heart-shaped.

SNEEZEWORT

Achillea ptarmica HEIGHT TO 60CM

Upright perennial plant with stiff, angular stems. It is widespread and locally common in much of Britain, favouring damp grassy places, including woodland rides, mainly on neutral to acid soils. It sometimes grows around the margins of rural gardens, where soil conditions suit its needs. Sneezewort flowers are borne in open clusters of heads, each 1–2cm across, with greenish-yellow inner florets and white outer florets; they appear July–September. Its leaves are narrow and untoothed.

FEVERFEW

Tanacetum parthenium HEIGHT TO 50CM

Branched, downy and aromatic perennial that has its origins in the garden; it is now widely naturalised in the countryside at large, being found growing in disturbed ground, and on verges and old walls. It is a cottage garden favourite and few old-style herbaceous borders are without it. Feverfew flowers are borne in open clusters of Daisy-like heads, 1–2cm across, with yellow central florets and white outer florets; they appear July–August. Its leaves are divided.

WILDFLOWERS

Enlightened wildlife gardeners adopt a tolerant approach to native wildflowers, tolerating species they might previously have considered weeds in at least part of the garden. As the horticultural adage says, 'a weed is just a plant growing in the wrong place'.

OXEYE DAISY

Leucanthemum vulgare
HEIGHT TO 60CM

Downy or hairless perennial that is widespread and common throughout, growing in dry, grassy places, often appearing in great quantity on recently disturbed ground. It is popular in the garden for its showy display and is a cottage garden favourite. Oxeye Daisy flowers are borne in heads, 30-50mm across, with yellow central florets and white outer florets; they appear May–September. Its leaves are dark green and toothed; the lower spoon-shaped leaves form a rosette, while stem leaves are lobed.

WINTER HELIOTROPE

Petasites fragrans
HEIGHT TO 20CM

Tolerant of frost, Winter Heliotrope flowers early and is a harbinger of spring. It is a patch-forming perennial garden plant that is now widely naturalised in hedgerows and on roadside verges; the extent to which it spreads is sometimes alarming since it excludes most other, native, species. The vanilla-scented flowers are borne in spikes of pinkish-lilac heads, each 10-12mm across; they appear December–March. Its leaves are rounded, 20cm across, long-stalked and can be found all year round.

RAGWORT

Senecio jacobaea HEIGHT TO 1M

Often reviled because it is poisonous to grazing animals (although the living plant is invariably avoided), Ragwort has a bad name in some quarters. In its favour it is a fantastic nectar source for insects and its leaves are eaten by Cinnabar Moth caterpillars. It is widespread and common in most parts. Ragwort flowers are borne in flat-topped clusters of yellow heads, each 15-25mm across; they appear June–November. The leaves are deeply divided with a blunt end lobe.

COLTSFOOT

Tussilago farfara HEIGHT TO 10CM

Charming, creeping perennial plant that appears in early spring as flowers carried atop leafless stalks, with purplish bracts. It grows on heavy, often waterlogged soils, and thrives on disturbance: it sometimes appears in great abundance on heavily disturbed ground, only to disappear in the years that follow as other plants colonise and stabilise the soil. Coltsfoot flowers are borne in heads, 15-35mm across, of yellow florets, and appear February–April and are popular with early spring insects. The leaves, which are rounded, heart-shaped and 10-20cm across, appear after flowering has finished and often carpet the ground in which the plant is growing.

Coltsfoot

LEFT: **Ragwort**

GROUNDSEL

Senecio vulgaris HEIGHT TO 40CM

A plant of recently disturbed ground, Groundsel is a familiar sight in the garden, often thriving in the tilled soil of the vegetable patch. It is widespread and common throughout Britain and in the countryside it is commonest in farmland. Groundsel flowers comprise cylindrical heads, 10mm long, with yellow florets surrounded by black-tipped greenish bracts; they appear January–December. The leaves are deeply divided; the lower ones are stalked while the upper ones are unstalked and clasp the stem.

GREATER BURDOCK

Arctium lappa HEIGHT TO 1M

Robust and downy plant of hedgerows, woodland rides, verges and waste ground; it sometimes appears on disturbed ground in the garden. Greater Burdock flowers are borne in egg-shaped heads, 20-40mm across, with purplish florets and greenish-yellow, hooked and spiny bracts; they appear July–September. Its fruits are burs, armed with hooked spines (flower bracts) that cling to animal fur and aid dispersal. The leaves are heart-shaped with solid stalks, and the basal leaves are longer than wide.

MARSH THISTLE

Cirsium palustre HEIGHT TO 1.5M

Upright, branched biennial that is often tinged reddish by late summer; its stems have continuous spiny wings. Marsh Thistle is common and widespread throughout Britain and grows in damp grassland; it sometimes grows in damp areas in gardens. Marsh Thistle flowers are borne in heads, 10-15mm across, with dark reddish-purple florets; the heads appear in clusters, July-September, and are extremely popular with butterflies and other insects. Its leaves are deeply divided, lobed and spiny.

PERENNIAL SOWTHISTLE

Sonchus arvensis HEIGHT TO 2M

Impressive perennial plant whose hollow stems yield a milky sap if broken; it grows in damp, grassy places and disturbed ground, and sometimes appears in garden borders. Perennial Sowthistle flowers are borne in umbel-like clusters of heads, each 4-5cm across, with yellow florets; they appear July-September. The fruits have feathery hairs (for wind dispersal) and form a 'clock'. Its narrow leaves have lobes, marginal spines and rounded clasping bases.

CAT'S-EAR

Hypochaeris radicata HEIGHT TO 50CM

Tufted perennial with hairless stems. It is common and widespread in dry grassy places and often grows on dry banks, and beside paths, in the garden. Common Cat's-ear flowers are borne in solitary heads, 25-40mm across, with numerous yellow florets; the flower stalks branch 1-2 times, are swollen beneath heads, and appear June-September. Its fruits are beaked with feathery hairs (to aid wind dispersal); the leaves are oblong, bristly and form a basal rosette.

YELLOW IRIS

Iris pseudacorus HEIGHT TO 1M

Familiar and robust plant that is common and widespread throughout Britain, growing in marshes and pond margins, and on riverbanks. It is often planted in the garden and grows well so long as the ground is wet enough. Yellow Iris flowers are 8-10cm across and bright yellow with faint purplish veins; they are borne in clusters of 2-3 flowers and appear May-August. Its fruits are oblong and 3-sided and the leaves are grey-green, sword shaped and often wrinkled.

SMOOTH SOWTHISTLE

Sonchus oleraceus HEIGHT TO 1M

Robust, hairless annual or biennial plant whose hollow stems yield a milky sap if broken. It is common and widespread in disturbed grassy places and often grows in gardens, favouring herbaceous borders and flowerbeds. Smooth Sowthistle flowers are borne in clusters of heads, each 20-25mm across, with pale yellow florets; they appear May-October. The fruits have feathery hairs to aid wind dispersal. Leaves have spiny margins and clasping bases.

COMMON DANDELION

Taraxacum officinale HEIGHT TO 35CM

Variable herbaceous plant of grassy places, with a taproot and a basal rosette of leaves. It is widespread and common in a wide range of grassy places, including gardens. Common Dandelion flower heads comprise a mass of yellow florets and are borne on hollow stems that yield a milky latex when broken; flowers appear April-September. The fruits have a hairy 'parachute', arranged as a white 'clock'; these aid wind dispersal. The leaves are spoon-shaped.

SNOWDROP

Galanthus nivalis HEIGHT TO 25CM

Familiar spring flower that grows in damp woodland where it often looks natural; however, it is probably naturalised or has been deliberately planted in most locations. Popular in the garden, it forms extensive patches where shady conditions match its favoured woodland setting. Snowdrop flowers are 15-25cm long and nodding, the 3 outer segments pure white, the inner 3 white with a green patch; they are solitary and nodding, and appear January-March. Its leaves are grey-green, narrow and all basal.

LORDS-AND-LADIES

Arum maculatum HEIGHT TO 50CM

Distinctive, familiar and welcome sight in spring, growing in hedgerows and woodlands and commonest in the south. It often grows in rural gardens and is often encouraged for its striking flowers and colourful berries. Lords-and-ladies flowers comprise a pale green, purple-margined spathe, cowl-shaped and part-shrouding the club-shaped, purplish-brown spadix; they are borne on slender stalks and appear April-May. Its fruits are red berries, borne in a spike; the leaves are arrowhead-shaped, sometimes dark-spotted.

ABOVE: **fruits**
LEFT: **flowers**

A surprising number of mammals are regular visitors to gardens, especially rural ones, some becoming surprisingly bold where encouraged by their owners. And these days a few notable species do far better in the company of man than they do in the countryside at large.

HEDGEHOG

Erinaceus europaeus
LENGTH 23-27CM

Instantly recognisable, the Hedgehog is now more common in suburban areas than in the surrounding countryside. Individuals may range widely during their nocturnal wanderings, taking a range of invertebrates including beetles, earthworms, slugs, snails and caterpillars. Hedgehogs normally enter hibernation during October, emerging again from March onwards. Some wake in midwinter and move hibernation sites. The population maybe in decline.

MOLE

Talpa europaea LENGTH 14-18CM

The Mole is perceived to be a pest of agricultural land and amenity grassland, including lawns, where molehills are all too evident for many gardeners. These are created during excavation of permanent tunnel systems used by the Moles to harvest invertebrate prey. Since research has shown 'sonic' Mole scarers to be totally ineffective, and because damage to lawns is largely superficial, it would seem better to welcome the Mole as a sign of a healthy soil than attempt to remove it.

ABOVE: **Hedgehog**; BELOW: **Mole**

PYGMY SHREW

Sorex minutus LENGTH 7-10CM

The diminutive Pygmy Shrew, weighing less than 6g, is our smallest mammal. It is similar in appearance to the Common Shrew but can be separated by the combination of very small size and proportionally longer, thicker and more hairy tail. Although they will use the burrows of other small mammal species, Pygmy Shrews spend less time underground than Common Shrews. They are also less vocal and do not produce the audible twitters characteristic of foraging Common Shrews.

COMMON SHREW

Sorex araneus LENGTH 9-14CM

Common Shrews are more often heard than seen; the twittering high-pitched calls are audible to most listeners and are usually produced while the shrew is foraging within the preferred thick grass or woodland habitats. The fur on the back is brown, contrasting sharply with the pale brown flanks and grey underbody. The small eyes and ears, together with the pointed face help to separate this (and our other shrews) from mice and voles. Spends a lot of time foraging below the surface.

BANK VOLE

Myodes glareolus LENGTH 13-17CM

With small eyes, reduced ears and a blunt face this is a typical vole. While adults show a rich chestnut-brown back with off-white underparts, immature animals are grey-brown in colour. The preferred habitat appears to be deciduous woodland, though many others are also used, including gardens, where thick ground cover is available. Bank Voles are active throughout the day, but with peaks of activity at dawn and dusk. They forage mostly on the ground but they will climb after hedgerow fruits.

WOOD MOUSE

Apodemus sylvaticus
LENGTH 15-22CM

With its large eyes, large ears and long tail, the Wood Mouse (also known as the Long-tailed Field Mouse) should be one of our most familiar mammals. An opportunistic and highly adaptable nature has enabled the Wood Mouse to occupy a very wide range of habitats (including urban gardens). In fact, you are more likely to encounter this species in your garden shed or house than a House Mouse. Note the large ears, big eyes and long tail.

YELLOW-NECKED MOUSE

Apodemus flavicollis
LENGTH 18-25CM

Large and more aggressive than the Wood Mouse, this species shows a restricted distribution, extending south of a line drawn from the Wash to the Welsh borders. Mature deciduous woodland is the preferred habitat but populations sometimes occur in hedgerows and rural gardens. The tail is usually longer than the body and is proportionally thicker at the base than seen in Wood Mouse. The best feature for identification, however, is the complete yellow throat collar, discernible even in juveniles as a grey band.

HOUSE MOUSE

Mus domesticus LENGTH 14-19CM

The House Mouse is a dull grey-brown in colour (compare this with the richer brown of Wood Mouse) and, for a mouse, has rather small eyes and ears. Within Britain, the House Mouse tends to live a commensal existence – alongside man – favouring houses, warehouses, shops and farm buildings. Because the young are sexually mature at 6 weeks of age, and because litters can be produced at monthly intervals, populations have the potential to increase rapidly if sufficient food is available.

Yellow-necked Mouse

COMMON RAT

Rattus norvegicus
LENGTH 30-50CM

Also known as the Brown Rat, this species was first introduced into Britain from Russia in the 1720s. Since then it has spread across most of the country, occupying a range of habitats, and by doing so it has displaced the Ship Rat. Rats, of both species, are perceived as pests, contaminating stored foodstuffs, passing disease and predating native wildlife. The large size, long scaly tail and rather pointed muzzle allow separation from mice and voles.

RED SQUIRREL

Sciurus vulgaris LENGTH 35-45CM

The native Red Squirrel is now largely restricted to Northern England, parts of Scotland, Anglesey and the Isle of Wight. Isolated populations may still be present in Yorkshire, Durham, Wales and Lancashire. The variable fur is uniformly dark, ranging from a rich chestnut-brown to a deep red-brown. Ear tufts are present in late summer and are also useful for identification. Red Squirrels make use of garden feeding stations where these happen to be in suitable habitat within the existing range.

GREY SQUIRREL

Sciurus carolinensis
LENGTH 45-55CM

Larger than our native species, the Grey Squirrel is predominantly grey in colour. However, some individuals may show a warm brown on their back and flanks, while others may be black or completely white. Grey Squirrels are resourceful creatures and quickly learn to exploit the resources available at garden feeding stations, working hard to find ways around deterrents set up to prevent them reaching bird feeders. Greys are found across most of England and along the central belt in Scotland.

RABBIT

Oryctolagus cuniculus
LENGTH 40-55CM

This species favours areas of short grassland (for feeding) alongside hedgerows or woodland where burrows can be established. Gardens may be visited, especially rural ones offering a range of succulent plant material. The Rabbit is smaller in size than a Brown Hare and also lacks the black tips to the ears that are characteristic of the latter species. While a hare will run some distance showing the black top to its tail, a Rabbit will bolt for its burrow with the white undertail showing.

ABOVE LEFT: Red Squirrel
LEFT: Grey Squirrel
BELOW: Rabbit

MAMMALS AND REPTILES

Mammals the size of deer and Badgers are regular visitors to many gardens but their nocturnal habits mean they are often overlooked, as are bats. More visible, because of their diurnal behaviour, are reptiles and gardens are important refuges for three species.

Muntjac

FOX

Vulpes vulpes LENGTH 95-130CM

With its erect ears, slender muzzle and bushy tail, the Fox should be instantly recognisable. This versatile species is virtually ubiquitous in Britain, occurring in a wide range of habitats and with distinct urban populations. Gardens are used regularly as is the food that is deliberately left out for their benefit by many householders. Family groups may breed in larger gardens, digging an earth in a bank or under the cover provided by a shed or outbuilding.

BADGER

Meles meles LENGTH 65-80CM

With its iconic white-striped face, the Badger is instantly recognisable. The species is found across most of lowland Britain, though it favours areas with deciduous woodland alongside earthworm-rich pasture. Setts are sometimes established in urban areas and Badgers may visit gardens to feed, sometimes digging up lawns for earthworms. Badgers are still persecuted in some areas, despite receiving legal protection; the threat of mass culls in England has been averted.

MUNTJAC

Muntiacus reevesi
SHOULDER HEIGHT 38-45CM

This dog-sized, reddish-brown deer, was first introduced to Britain from southeast China and Taiwan. It has now spread and can be found across much of the country, with core populations centred on the east Midlands and East Anglia. Males have small simple antlers and characteristic blank stripes running up the ginger face. Although they prefer thick cover, Muntjac will visit gardens, even suburban ones, where they are close to areas of suitable habitat.

ROE DEER

Capreolus capreolus
SHOULDER HEIGHT 65-70CM

The Roe is a medium-sized deer with long legs, black nose and white-chin. When disturbed, a Roe will run off with a characteristic bounding gait, displaying the creamy-white rump patch and appearing tailless. Roe are found in woodland and nearby farmland habitats across most of Scotland, northern England, East Anglia and the south of England. They may be found in suburban or even urban situations, such as parks and golf courses, if sufficient cover is available.

STOAT

Mustela erminea LENGTH 25-45CM

Despite its long thin body and short legs, the Stoat is a surprisingly quick and agile predator. Larger in size than a Weasel, a full-grown Stoat is quite capable of tackling a Rabbit. The tail, which is longer than that seen in a Weasel, is tipped with black. A few Stoats develop a white coat in winter. Stoats rarely venture into gardens but may visit those of a rural nature in the search for small mammals and birds.

WEASEL

Mustela nivalis LENGTH 20-40CM

This is our smallest carnivore; its long, thin body and short legs are suited for the pursuit of small mammals in their burrow systems. The small size and short tail (lacking a black tip) help separate this species from the larger Stoat to which it is related. Weasels tend to occupy the same habitats as their small mammal prey, favouring hedgerows and other linear features along which they hunt. Birds and eggs are also taken occasionally.

Roe Deer

Stoat

Weasel

BROWN LONG-EARED BAT

Plecotus auritus
WINGSPAN 24-28CM

This medium-sized bat can be distinguished from all others, except the very similar Grey Long-eared Bat, by the very large ears. The species is found across all except the most northerly parts of Britain, roosting mainly in suburban and rural buildings close to woodland throughout the summer months. During the winter, the bats use tree cavities, caves, mine shafts and outbuildings for hibernation. Brown Long-eared Bats feed on a range of moths, beetles and flies, with prey items typically bigger than 3mm in size.

TOP: **Brown Long-eared Bat**
ABOVE: **Common Pipistrelle**

COMMON PIPISTRELLE

Pipistrellus pipistrellus and
Soprano Pipistrelle *P. pygmaeus*
WINGSPAN 18-24CM (BOTH SPECIES)

Although the Pipistrelle has long been regarded as one of our most familiar bat species, it was recently discovered that the pipistrelle population in Britain actually comprised two different species – the Common Pipistrelle and the Soprano Pipistrelle. Researchers discovered that the two species produced slightly different echolocation calls, behave in somewhat different ways and look subtly different in the hand. But they are still struggling to work out how the two species are able to co-exist without competing for food.

SLOW-WORM

Anguis fragilis
LENGTH 3-40CM

This legless lizard may be encountered across much of Britain but is most abundant in the warmer southern counties. Slow-worms are relatively easy to spot and seem to like living alongside Man. For this reason, they are commonly reported from gardens, allotments and urban parkland. Slow-worms spend a lot of time foraging in deep cover or underground and may be discovered under logs or buried in compost heaps. They are a uniform shiny grey-brown in colour, with older males sometimes showing a pattern of blue spots.

Brown Long-eared Bat

Common Pipistrelle

Soprano Pipistrelle

GREY LONG-EARED BAT
Plecotus austriacus
WINGSPAN 26-30CM

This species is confined to the extreme south of Britain, roosting in the roofspace of houses in open wooded country. Hibernation takes place in cellars, caves and old brick kilns. Old houses and churches are often favoured in summer, the bats roosting on the ridge beam. The Grey Long-eared Bat seems to catch more of its prey in open flight, rather than gleaning it from the surface of leaves as preferred by its more common relative.

COMMON LIZARD

Lacerta vivipara LENGTH 10-15CM

The Common Lizard is well adapted to living under cool conditions – a behaviour that has allowed it to occupy a range of habitats across Britain. While damp heathland is preferred, it can also be found in woodland, on brownfield sites and in larger gardens. Common Lizards are variable in their coloration but any lizard with legs seen in a garden is likely to be a Common Lizard. The other native species, the Sand Lizard, is rare and restricted to heathland.

GRASS SNAKE

Natrix natrix LENGTH 60-130CM

The Grass Snake is the only one of our snakes likely to be encountered in a garden. Damp areas, with long grass and close to waterbodies, are preferred and eggs are sometimes laid in compost heaps. Adults may reach a metre in length but young Grass Snakes are about the size (and thickness) of a pencil. They are usually a medium olive green above, with a regular pattern of black blotches along their length and a yellow and black 'collar' behind the head. The Grass Snake is not poisonous.

Slow-worm

Grass Snake

Grass Snake

Common Lizard

AMPHIBIANS & BUTTERFLIES

With the countryside under threat from development and farming, gardens are increasingly important refuges for our native amphibians. Garden ponds are the key to attracting these species while an abundance of nectar-rich flowers and a tolerance of native plants helps butterflies.

SMOOTH NEWT

Triturus vulgaris LENGTH 9-10CM

This is the most common of our newts and is found throughout Britain; it is the only newt in Ireland. A range of ponds (free from fish) is used, the species preferring those waters that are not too acidic. During that part of the year when the Smooth Newt lives out of water, it may be found under rockery stones and amongst garden debris. Males have a breeding season crest and both sexes have spots on the throat and along the belly.

PALMATE NEWT

Triturus helveticus
LENGTH 8-9CM

With a more restricted and westerly distribution than Smooth Newt, the Palmate Newt prefers shallow ponds and pools with more acidic water (e.g. heathland and woodland ponds). Males develop a low crest in the breeding season. The tail ends abruptly and is tipped by a short, hair-like, filament; the hind feet are fully webbed and the throat unspotted, all features that allow separation from Smooth Newt. A dark stripe is present on the side of the head.

GREAT CRESTED NEWT

Triturus cristatus LENGTH 14-16CM

This is our largest newt and, with its rough warty skin, it is readily identifiable. The underside is brightly coloured with orange or yellow, interspersed with a series of dark blotches, the pattern of which is as unique as a fingerprint. This species is more aquatic in habits than the other newts and prefers deeper pools with plenty of aquatic vegetation. Adults arrive back at their breeding ponds February–April. This newt Is protected by law.

COMMON FROG

Rana temporaria LENGTH 6-10CM

The Common Frog can be found right across Britain, occupying a range of habitats, and may be found at some distance from water when foraging. Garden ponds have become increasingly important for the species, providing new opportunities for breeding but, like other waterbodies, are only used for a small part of the year. The species is normally brown or olive green but unusual colour forms are sometimes reported, including red or orange individuals.

COMMON TOAD

Bufo bufo LENGTH 5-9CM

Although the Common Toad has a similar distribution to the Common Frog, it is less abundant over its range and is more restricted in its habitat preferences. It is thought that Common Toad tadpoles are out competed by those of the Common Frog. This may be why the species shows a preference for larger ponds containing fish. Toad tadpoles are poisonous and are avoided by many predators, while frog tadpoles are readily taken.

LARGE WHITE

Pieris brassicae WINGSPAN 60MM

The Large White is found throughout Britain and Ireland, and has 2 generations each year; the first is on the wing from April, the second (more numerous) generation is on the wing from July. As well as being larger than our other whites, it can be identified by the blacker wingtips and bolder spots. The Large White is sometimes unwelcome in gardens because of the damage its caterpillars inflict on cabbages and other brassicas.

SMALL WHITE

Pieris rapae WINGSPAN 45MM

Although sometimes lumped with the Large White under the name 'cabbage white', this species does far less damage to brassicas. Smaller in size than the Large White, it may be confused with both Green-veined White and female Orange Tip. First brood individuals (April–May) have pale black tips to the wings and less strongly marked spots. Individuals from the second brood (August) are larger and more strongly marked than those of the first brood.

ORANGE-TIP

Anthocharis cardamines WINGSPAN 40MM

Orange-tips are on the wing April-July, males instantly recognisable and a sure sign of spring. Females have black (not orange) wingtips and single black spot on the upperwing; the underwing pattern is useful for identification: white is the base colour, with a pattern of green camouflage across it. Orange-tips are commonly seen in gardens and may lay their eggs on Honesty and Dame's Violet; Cuckooflower is the typical native larval foodplant.

CLOUDED YELLOW

Colias croceus WINGSPAN 50MM

The Clouded Yellow is an uncommon summer visitor to southern Britain and Ireland, arriving from southern Europe in varying numbers each year. Males have distinctive marigold-yellow (almost orange) upperparts, the wings edged with a solid black band. Females share this band but have upperparts that are a pale greenish white, smudged with black. Migrants are common on southern coasts and smaller numbers spread inland and further north. Larval foodplants include Lucerne.

PAINTED LADY

Vanessa cardui WINGSPAN 60MM

This familiar species can be found across most of Britain, though it is more abundant in the southern half reflecting its migratory status. Individuals from populations in North Africa and the Middle East migrate to Britain in most years, bringing them to garden nectar sources from May onwards. The species recalls the Small Tortoiseshell but is still readily identifiable in the field. Painted Lady larvae feed on thistles but pupae cannot survive our winters.

ABOVE:
upperwing
LEFT:
underwing

GREEN-VEINED WHITE

Pieris napi WINGSPAN 45-50MM

Although this widely distributed species is frequently seen in gardens, it actually prefers damper habitats such as woodland rides and shady margins of moist meadows. Difficult to identify in flight, the Green-veined White is best identified at rest, when the dusky dark smudging along the wing veins can be seen. The species is variably dark-dusted along the veins. Two or sometimes 3 generations are on the wing April-October. Larval foodplants include Garlic Mustard.

BRIMSTONE

Gonepteryx rhamni WINGSPAN 60MM

The Brimstone overwinters as an adult and so is one of the first butterflies to be seen in spring. The sulphur-yellow coloured males are quite conspicuous but the pale females, which are greenish white in colour, may be confused with a Large White. The Brimstone is found across southern Britain and Ireland, extending as far north as the Pennines. Brimstones always rest with their wings folded above the back. Larval foodplants are Buckthorn and Alder Buckthorn.

SMALL TORTOISESHELL

Nymphalis urticae WINGSPAN 45MM

Because the Small Tortoiseshell overwinters as an adult, individuals may be encountered in all months of the year. The size of the population varies from year to year, often dramatically, and large numbers may be seen in the autumn, visiting garden Butterfly-bush flowers to feed on nectar prior to entering hibernation. Individuals may be found hibernating in sheds and other outbuildings. The larvae feed on Common Nettle.

RED ADMIRAL

Vanessa atalanta WINGSPAN 60MM

With its dark brown (almost black) and red upperparts, this striking species is readily identifiable. Like the Painted Lady, numbers seen here in summer are determined by the scale of arrivals from southern Europe and North Africa. However, recent mild winters have allowed the species to overwinter successfully and soon it may be partly resident in Britain. Arrivals peak in June, then individuals drift further north, before retreating south again in August. The larvae feed on Common Nettle.

LEFT:
upperwing
BELOW:
underwing

Butterflies are a welcome addition to any garden. Some are visitors from the surrounding countryside that come in search of nectar but a surprising range of species actually breed, as well as feed, in the garden where thoughtful owners grow and encourage their larval foodplants.

PEACOCK

Nymphalis io WINGSPAN 60MM

The bold eye-spots, used to deter would-be predators, help give this species its name. They also make the Peacock easy to identify. Its widespread distribution (all but the extreme north of Britain) make the Peacock a familiar garden visitor. After hibernation, adults emerge February-March but this butterfly is most abundant in July, when the single annual generation emerges. The larval foodplant is Common Nettle, the species preferring those in sunny positions within the middle of a larger patch.

COMMA

Polygonia c-album
WINGSPAN 45MM

The Comma's ragged outline is unique amongst British butterflies and is thought to help camouflage the butterfly as it hibernates among dead leaves at the base of a tree. The Comma has been expanding its southerly distribution in recent years, pushing further north into southern Scotland. Now that it is so widespread it is strange to think that it was once nearly extinct here. Males establish territories in gardens. Larval foodplants include Common Nettle and Hop.

LEFT: **pupa**
BELOW: **upperwing**

TOP: **upperwing**
ABOVE: **underwing**

SPECKLED WOOD

Pararge aegeria WINGSPAN 45MM

Although this markedly shade-tolerant species is a butterfly of woodland rides and clearings it will venture into gardens and establish breeding territories there. The two sexes are similar in appearance, with pale blotches on a darker brown background. Small black eye spots are present within some of the paler blotches, those of the female typically larger in size than seen in the male. The species is double-brooded and on the wing March-October. The larvae feed on grasses.

ABOVE:
upperwing
LEFT:
underwing

GATEKEEPER

Pyronia tithonus WINGSPAN 40MM

As the alternative name of Hedge Brown suggests, this is a butterfly of hedgerows and country lanes. It is also found in a variety of habitats where there is an abundance of tall vegetation and shelter. As such, it can be found in more rural gardens, often exploiting the nectar sources available there and visiting from neighbouring farmland. It has a southerly distribution, extending north to Lancashire and Yorkshire and west to southern Ireland. It flies July-August; the larvae feed on grasses.

ABOVE:
upperwing
LEFT:
underwing

RINGLET

Aphantopus hyperantus
WINGSPAN 48MM

The Ringlet favours damp woodland and shady hedgerows; its range extends into central Scotland but it is commonest in southern England. A recent expansion in range has also seen the Ringlet reoccupy suburban sites from which it was lost some time ago. It has very dark brown upperwings marked with dark eye-spots, faintly edged with pale brown. Those of the female are more obvious than seen in the male. It flies June-July; the larvae feed on grasses.

ABOVE:
upperwing
LEFT:
underwing

MEADOW BROWN

Maniola jurtina WINGSPAN 50MM

The Meadow Brown lays its eggs on rank grasses and is thus able to utilise a range of grassland habitats across its wide British range. There is a single generation each year, on the wing May-September. Males are uniformly dark above, much like a Ringlet but with a single rust-brown edged black eye-spot on each forewing. Females look a bit like a large Gatekeeper but have a muddier ground colour and only a single white dot to each eye-spot.

ABOVE:
upperwing
LEFT:
underwing

SMALL HEATH
Coenonympha pamphilus
WINGSPAN 30MM

A common grassland butterfly but easily overlooked. It favours dry, light soils, supporting finer grasses (food for its larvae), so it is only an occasional garden visitor. The warm ochre ground colour of the upperwings tends to be paler in the larger female than in the male. A single eyespot is present on each forewing of both sexes, though this can appear very washed-out in the female. A double-brooded species, it flies May-June and August-September.

SMALL COPPER
Lycaena phlaeas WINGSPAN 25MM

This restless, mobile butterfly is found in woodland rides, waste ground and roadside embankments. It visits gardens to feed. The Small Copper thrives in warm, sunny conditions so hot summers boost the population while cool, wet seasons cause a decline. Both sexes have upperparts that are a mix of dark brown and bright coppery orange. With 2-3 broods each year, it flies May-September. The larvae feed on Sheep's Sorrel.

LEFT:
underwing
BELOW:
upperwing

HOLLY BLUE
Celastrina argiolus
WINGSPAN 30MM

The Holly Blue can be found in woodlands, shrubby parks and gardens, where it lays its eggs on Holly (spring brood) and Ivy (autumn brood). Both sexes show blue upperparts and pale blue underparts, those of the female with black wingtips. The forewings are finely edged black with narrow white margins, the margins broken up into sections by black lines extending from the blue out to the wing's edge. It flies April-May and August-September, in 2 broods.

ABOVE:
upperwing
LEFT:
underwing

LARGE SKIPPER
Ochlodes venatus
WINGSPAN 35MM

This species is widespread in Britain, extending north into southern Scotland, but not Ireland. The upperparts have a stronger orange-brown coloration than seen in the other two common skippers and also have a more pronounced pattern of darker markings. Grassland habitats are favoured; churchyards and urban parks with areas of long grass encourage the species into urban areas. Bramble flowers are visited for nectar. Flies June-July; larvae feed on grasses.

PURPLE HAIRSTREAK
Neozephyrus quercus
WINGSPAN 38MM

The Purple Hairstreak is a species of the oak canopy, feeding on the honeydew produced by canopy-dwelling aphids, and is only rarely seen at ground level. As such, only rural gardens, with mature oak nearby are likely to be visited. This stunning little purple and black butterfly is on the wing June-September. Its distribution is patchy but does reach as far north as central Scotland. The larvae feed on oaks.

ABOVE:
upperwing
LEFT:
underwing

COMMON BLUE
Polyommatus icarus
WINGSPAN 32MM

The Common Blue is less likely to visit gardens than the Holly Blue. It is mainly a butterfly of flowery meadows, heathland and woodland rides. The male sports blue upperparts, the female brown, and both sexes show grey-brown underparts, patterned with white, orange and black spots. Male's upperwings are edged black with a white margin. Flies April-September in several broods; larvae feed on trefoils.

ABOVE LEFT:
male
ABOVE:
female
LEFT:
underwing

SMALL SKIPPER
Thymelicus sylvestris
WINGSPAN 25MM

The Small Skipper is a common grassland butterfly whose distribution extends across much of England and Wales. The rusty orange upperparts are edged with black in both sexes, the male additional sporting a black line in the centre of the wing. The Small Skipper is similar to the Essex Skipper but can be identified by the orange-brown underside of the tips of the antennae (black in Essex Skipper). It flies July-August and the larvae feed on grasses.

Small Skipper
Essex Skipper

ESSEX SKIPPER
Thymelicus lineola
WINGSPAN 25MM

The Essex Skipper was not identified as a separate species to the Small Skipper until 1889. Its range is more restricted than its cousin, being found in southeast England and in East Anglia. The two species use the same habitats and fly at the same time (July-August). Both favour grassland habitats, including roadside verges, and may be seen within gardens. The Essex Skipper's British range is expanding. Its larvae feed on grasses.

Moths are an extremely diverse group of insects, most of which lead nocturnal lives as adults and hence are easily overlooked. But some species are large and spectacular enough to be discovered on a regular basis by keen-eyed gardeners. They can be encouraged by growing appropriate larval foodplants.

LIME HAWKMOTH

Mimas tiliae
FOREWING LENGTH 23-29MM

This species is found throughout England, north as far as Yorkshire and locally within Wales. The colour of the wings, together with their shape, help to distinguish the Lime Hawkmoth from all our other resident hawkmoths. On the wing May–July, this species may sometimes be found at rest on walls or tree trunks. The larva feeds on limes, elms, Silver Birch and Alder, and sometimes on Wild Cherry cultivars.

Lime Hawkmoth

Poplar Hawkmoth

Poplar Hawkmoth larva

POPLAR HAWKMOTH

Laothoe populi
FOREWING LENGTH 30-46MM

This common and widespread resident is perhaps the most frequently encountered hawkmoth, occurring across a range of habitats, including gardens. The caterpillars feed on poplars, Aspen, and various species of willow. When fully grown, the caterpillar is green in colour, with yellow stripes on its side. The Poplar Hawkmoth usually has just one generation per year, on the wing May–August. It is regularly attracted to light and may be frequently encountered in moth traps.

EYED HAWKMOTH

Smerinthus ocellata
FOREWING LENGTH 36-44MM

The use of foodplants such as willows and apples means that this species is commonly encountered across much of lowland England (though not Scotland). If its hind wings are visible then this species should be instantly recognisable, since each sports a clear eye-spot on a pink background. Eye-spots are used to scare away potential predators. The Eyed Hawkmoth is on the wing May–July and has a single generation each year.

ELEPHANT HAWKMOTH

Deilephila elpenor
FOREWING LENGTH 28-33MM

This species is widespread and generally common except in the far north. It favours a range of different habitats, including gardens, and the main larval food-plants are Rosebay Willowherb and Great Willowherb. It is on the wing May–August, often visiting Honeysuckle for nectar. The pinkish-olive forewing and pinkish-black hind wing make this attractive hawkmoth fairly easy to identify. The related Small Elephant Hawkmoth is smaller, differently marked and has more pink on the body.

larva

PRIVET HAWKMOTH

Sphinx ligustri
FOREWING LENGTH 41-55MM

The Privet Hawkmoth has a largely southerly distribution, becoming more local (and coastal) in its distribution as you move north towards Scotland. Superficially similar to the Pine Hawkmoth, it is the larger of the 2 species and has pink on its abdomen. The beautiful lime green caterpillar has purple and white stripes that run obliquely across its body and a hook-like appendage on its rear. The single generation is on the wing June–July, favouring gardens on chalky soils. The larvae feed on privets and lilacs.

Privet Hawkmoth larva

Elephant Hawkmoth

PINE HAWKMOTH
Hyloicus pinastri
FOREWING LENGTH 35-41MM

The spread of conifer plantations has benefited this rather local species and it is now more widely distributed than it once was. The main foodplant is Scots Pine but Norway Spruce and Cedar of Lebanon may be used. Honeysuckle is well used for nectar and the moth may be attracted to light, sometimes in large numbers. Adults are on the wing May–August and may be found by day resting on the trunks of trees or on fence posts.

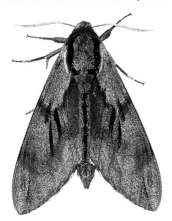

CURRANT CLEARWING
Synanthedon tipuliformis
FOREWING LENGTH 8-10MM

The group to which this small moth belongs is named because of the translucent areas on their wings. Currant Clearwing larvae feed inside the stems of currants (wild or cultivated) but may also use gooseberry. It is probably widespread in Britain but is under-recorded by observers. Although it flies in the daytime, it is easily overlooked, seldom straying far from its larval food-plants and is alert and furtive. Garden currant patches are worth checking in May and June.

PUSS MOTH
Cerura vinula
FOREWING LENGTH 29-38MM

Both this black and white moth and its clown-faced caterpillar are striking, the caterpillar all the more so for the strange appendages it raises when threatened. Widely distributed throughout most of the British Isles, the species is regularly encountered in gardens, areas of open woodland and near hedgerows. The larva feeds on poplars and willows, favouring those in sunny locations. The Puss Moth flies May–July and has 1 generation per year.

larva

CHOCOLATE-TIP
Colstera curtula
FOREWING LENGTH 13-18MM

This small moth has two generations each year, the first on the wing April–May, the second August–September. Aspen, together with various poplars and willows, are used by the caterpillar as foodplants. This preference sees the species favour areas of woodland, scrub and larger gardens in rural areas. Although well distributed across southern England it is scarce elsewhere, although there is an isolated population near Inverness.

HUMMINGBIRD HAWKMOTH
Macroglossum stellatarum
FOREWING LENGTH 20-24MM

This migrant species is mainly a late summer visitor to Britain, peaking in abundance in August. Visitors to Britain originate from populations in southern Europe and North Africa, arriving here in numbers that vary greatly from one year to the next. The species may be seen flying during the daytime and feeding at Butterfly-bush and other flowers. It is easy to see why some observers confuse this moth with a hummingbird, because it feeds in a similar manner.

LAPPET MOTH
Gastropacha quercifolia
FOREWING LENGTH 28-42MM

This characteristic moth is most commonly encountered in the extreme south of Britain, its distribution further north being rather patchy. It is thought that numbers have been in decline for some time. The favoured habitats are thorny scrub (the caterpillars feed on Blackthorn, Hawthorn and Crab Apple), hedgerows and rural gardens. The moth's resting posture aids identification, as does the large size, purple-brown colour and prominent snout. It flies June–August.

larva

BUFF-TIP
Phalera bucephala
FOREWING LENGTH 22-34MM

When viewed at rest this moth should be unmistakable and, with the wings held against the body, it resembles a broken birch twig. So effective is this disguise that the moth is easily overlooked. Although local in its Scottish distribution it is more widely distributed further south in England and Wales. Open woodland, scrubby habitats and gardens are all favoured, the caterpillar feeding on various tree species. The Buff-tip is on the wing May–July.

IRON PROMINENT
Notodonta dromedarius
FOREWING LENGTH 18-24MM

This beautiful moth is one of a number of species named after the prominent projections raised above the back when at rest. It is thought that these help to break up the outline of the resting moth, camouflaging it from predators. It occurs throughout Britain and flies May–June (first generation) and again July–August (second generation). There is only 1 generation per year in the north. Larval foodplants include birches, oaks, Alder and Hazel.

Leave an outside light on overnight, or better still invest in a moth trap, and you will be amazed at the number of moths that occur in even the smallest and most urban of gardens. Key to their success and survival in any garden setting is the presence of plenty of larval foodplants and places to pupate.

PALE TUSSOCK

Calliteara pudibunda
FOREWING LENGTH 21-31MM

The Pale Tussock is a common resident, found throughout England and Wales but seemingly absent from Scotland. It is a common species in gardens, including urban ones, and is on the wing May-June. Females are larger than the males and show a more elongated shape when at rest. The larvae feed on a wide variety of trees and shrubs and were extremely abundant on commercial Hops prior to the introduction of insecticides.

WHITE ERMINE

Spilosoma lubricipeda
FOREWING LENGTH 18-23MM

The small black dots on a white background give this moth its name 'ermine'. However, there is some variation in the tone of the white background coloration, with individuals in the southern half of Britain a crisp bright white and those in Scotland typically creamy buff. Gardens and other suburban and urban habitats are well used, perhaps reflecting the choice of larval foodplants: these include Common Nettle and docks. The adult flies May-July.

LESSER SWALLOW PROMINENT

Pheosia gnoma
FOREWING LENGTH 20-26MM

This species and the closely related Swallow Prominent are very similar in appearance and tricky to separate if inexperienced. Both are common and widely distributed, occupying woodland, parks and gardens as well as other habitats. The Lesser Swallow Prominent has 2 generations each year, the first on the wing from April-June, the other from July-August. The larvae feed on birches and the species overwinters as a pupa.

PALE PROMINENT

Pterostoma palpina
FOREWING LENGTH 18-25MM

This is a distinctive moth that resembles a fragment of broken wood. It is fairly common throughout the southern half of Britain, though becomes more local in northern England and Scotland. Northern populations have a single generation each year, on the wing May-June, but those in the south have 2. Feeding on various willows and poplars, this species is regularly encountered in gardens. The larva may be found on its foodplants throughout much of the summer.

Pale Tussock larva

CINNABAR

Tyria jacobaeae
FOREWING LENGTH 17-23MM

The Cinnabar is an unmistakable and familiar moth, commonly encountered in areas of grassland, open woodland and heathland. Its orange-and-black striped caterpillars feed on Ragwort and in some years may strip plants over quite an area. Although this moth may be attracted to light when flying at night, it is readily disturbed from vegetation during the day and may be seen on the wing in bright sunshine. The single generation is on the wing May-August.

GARDEN TIGER

Arctia caja
FOREWING LENGTH 28-37MM

The Garden Tiger is a stunning moth, with beautifully patterned brown and white forewings and striking orange hind wings, complete with blue spots. The hind wings are displayed if the moth is threatened, a behaviour sometimes accompanied by the production of a pungent clear yellow liquid from behind its head. This widespread but declining species favours a wide range of habitats including gardens. Flies July-August; larvae feed on a wide range of plants.

Garden Tiger

Cinnabar larva

HEART AND DART
Agrotis exclamationis
FOREWING LENGTH 15-19MM

Like many of its close relatives, this species is variable in its coloration. However, there are several consistent marks on the forewing that aid identification – one heart-shaped, another dart-like. The main flight period is May–August but sometimes a small second generation is seen in September. This moth comes readily to light and is often the most numerous species in moth traps. The larvae feed on a wide range of plants, both cultivated and wild, including Ribwort Plantain.

BROAD-BORDERED YELLOW UNDERWING
Noctua fimbriata
FOREWING LENGTH 22-27MM

This robust moth is found in wooded areas and regularly visits rural gardens. Its caterpillars feed on Common Nettle, docks, and a range of other herbaceous species. The moth flies July–September and is commonest in August and September. The forewing is variable in colour but consistent in pattern. This, coupled with the marigold yellow hind wing with its broad black border, should allow identification.

GREY DAGGER
Acronicta psi
FOREWING LENGTH 17-20MM

This attractive species has a soft grey ground colour to the wings and a striking dagger-like black mark near the forewing's outer margin. The colour and markings are similar to the closely related Dark Dagger (the two are hard to distinguish) although this species usually has a darker ground colour to the wings. The Grey Dagger is on the wing May–August, often visits mature gardens, and the larvae feed on a wide range of deciduous trees and shrubs.

ANGLE SHADES
Phlogophora meticulosa
FOREWING LENGTH 21-25MM

The creased way in which this moth folds its wings when resting makes it unmistakable. The Angle Shades may be encountered throughout the year but is most frequently observed May–October. A small peak in abundance occurs in May and June, but a larger one is often noted in August and September when numbers are boosted by the arrival of immigrants from further south. The species is common in gardens. Its larvae sometimes feed on potted pelargoniums.

HEBREW CHARACTER
Orthosia gothica
FOREWING LENGTH 15-17MM

Although the forewing colour of this widespread and common moth can range from sandy brown to almost black, the saddle-shaped mark on the wing is consistent and gives the moth its name. The Hebrew Character is a spring moth, on the wing March–May, often flying very late in the night to feed on sallow catkins. It is found in a wide range of habitats, from the smallest urban gardens to northern, upland moorland. It is perhaps commonest in lowland England.

LARGE YELLOW UNDERWING
Noctua pronuba
FOREWING LENGTH 21-26MM

This ubiquitous species is widespread across Britain and on the wing for a single protracted generation. Although the peak in activity appears to be in late August, the Large Yellow Underwing may be encountered June–November. Numbers here may be boosted by the arrival of immigrants from further south, with egg-carrying immigrants often arriving here before resident females have developed to this stage. The larvae eat a wide range of plants, including marigolds, docks and several grasses.

OLD LADY
Mormo maura
FOREWING LENGTH 30-36MM

This large dark moth is one of a number of species with intriguing names. It has a single generation each year; adults are on the wing July–September and are found in gardens and woodlands, as well as alongside riverbanks and areas of wet grassland; it sometimes ventures indoors to roost in the daytime. The Old Lady is well distributed throughout much of Britain but is commonest in the south. The larva feeds on various chickweeds, docks, Ivy and Hawthorn.

GREEN SILVER-LINES
Bena bicolorana
FOREWING LENGTH 19-23MM

The bright green, broad forewings, with their two parallel diagonal silvery-white lines, make this species instantly recognisable. On the wing June–July, the Green Silver-lines favours broadleaved woodland and parkland, sometimes venturing into the more rural gardens. Although fairly frequently observed in the southern half of Britain, it becomes more localised north of the Humber. The larva feeds on the buds and leaves of oak.

Many of the moths that are found in the garden are attractive to look at, with wings that are beautifully patterned, or colourful, or both. The markings are not for our benefit but rather they serve in the main to camouflage the resting insect from the eyes of predators such as birds.

BURNISHED BRASS

Diachrysia chrysitis
FOREWING LENGTH 16-19MM

The bold, brassy yellow markings on this moth are striking and often metallic in appearance, The Burnished Brass is found in gardens, along ditch banks and woodland edge. Its distribution, widespread throughout Britain, reflects the choice of larval foodplants; these include Common Nettle and White Dead-nettle, both of which are abundant. There is usually a single generation each year, on the wing June-July, but southern populations may have a partial second generation August-September.

HERALD

Scoliopteryx libatrix
FOREWING LENGTH 19-23MM

With its scalloped forewings and rich tones, the Herald is easily separated from other moths. Unlike many species the Herald overwinters as an adult, seeking out a sheltered outbuilding or cavity in which to sit out the winter (the numbers gathering at suitable sites may be considerable). These make up the single annual generation and are on the wing August-November, and again March-June. They favour open woodland habitats (including gardens) where willows and poplars are found.

ABOVE: **Burnished Brass**; BELOW: **Herald**

SILVER Y

Autographa gamma
FOREWING LENGTH 13-21MM

The Silver Y is a familiar immigrant, recorded throughout the year but most commonly May-August. It flies in sunny weather as well as after dark, nectaring at flowers alongside other immigrants (like Painted Lady). Although this is just one of several species to show a 'y' mark on each forewing, this is the one that is most often encountered. Immigrant Silver Y moths may breed here, the larvae eating Common Nettle, bedstraws, peas and runner beans.

RED UNDERWING

Catocala nupta
FOREWING LENGTH 33-40MM

This large moth has grey forewings but the hind wings – exposed when the moth feels threatened – are red with a broad black border and a thinner, centrally placed, black band. The Red Underwing rests on walls, fences and tree trunks by day and is attracted to light at night. It flies August-October and overwinters as an egg, deposited in a crack on a willow or poplar trunk. Woodland, scrub and gardens are used across its southerly range.

BELOW: **Red Underwing**

SVENSSON'S COPPER UNDERWING

Amphipyra berbera
FOREWING LENGTH 21-26MM

This species and its similar cousin the Copper Underwing *A. pyramidea* favour wooded habitats and are both regular visitors to mature gardens, venturing into sheds and indoors in the daytime. They are on the wing July-September and their larvae feed on various trees including oaks. Similar in appearance, they are best told apart by subtle differences in patterns on their coppery hind wings.

BLOTCHED EMERALD

Comibaena bajularia
FOREWING LENGTH 14-17MM

As you might guess from its name, this moth has soft green forewings marked with a number of paler lines and a series of fawn blotches around the margins. It flies June-July and males come more readily to light than females. Mature oak woodland is the favoured habitat - the larvae feed on oaks and Hazel - but individuals may be found in nearby gardens. The distribution is centred on southern England.

GREEN CARPET

Colostygia pectinataria
FOREWING LENGTH 12-15MM

Freshly emerged individuals are a beautiful green colour with a distinctive pattern of cross lines and darker blotches. The colour fades with age, but, once learned, the pattern is diagnostic. It provides camouflage on lichens on tree bark. The Green Carpet is a common and widespread resident, on the wing June-August in the north of its range. Two generations occur in the south, the first on the wing May-July, the second August- September. The larva feeds on bedstraws.

MAGPIE

Abraxas grossulariata
FOREWING LENGTH 18-25MM

This striking moth, with its white, black and orange-yellow wings, should be familiar to many. It is well distributed throughout Britain but occurs at relatively low densities in many areas. This may reflect the fact that it seems to be less common now than it was several decades ago. The northern heather moorlands support the greatest numbers but it is also found in woodland and gardens. It flies June-August and the larvae feed on various shrubs.

LILAC BEAUTY

Apeira syringaria
FOREWING LENGTH 19-22MM

This medium-sized moth rests with its forewings slightly raised and the leading edge creased. This, coupled with the lilac markings and bold line running across the forewing, confirm identification. The Lilac Beauty is primarily a moth of broadleaved woodland, though venturing into some gardens, and is found across England, north into the southern parts of Scotland. It is on the wing June-July and the larvae feed on Honeysuckle and privets.

BRIMSTONE MOTH

Opisthograptis luteolata
FOREWING LENGTH 14-21MM

This yellow moth is unmistakable, with its brimstone base colour and pattern of chestnut-brown markings. The Brimstone Moth is commonly encountered across its wide British range, being found in woodland, gardens, on heathland and grassland. Blackthorn and our two hawthorn species are larval foodplants. There are 2 or 3 generations each year, on the wing April-October. It sometimes flies in the late afternoon and comes to light at night.

SILVER-GROUND CARPET

Xanthorhoe montanata
FOREWING LENGTH 14-17MM

The Silver-ground Carpet favours damp areas with lush vegetation and occupies woodland rides, fens and some gardens. It is a rather common species, resident and distributed throughout Britain. A separate subspecies is found on Shetland. The caterpillars feed on a range of herbaceous plants, including Hedge Bedstraw and Cleavers. It comes to light and is often numerous at dusk but it is also disturbed from vegetation in the day. It flies May-July.

CANARY-SHOULDERED THORN

Ennomos alniaria
FOREWING LENGTH 16-20MM

The Canary-shouldered Thorn is one of a group of related species that have yellow or ochre-coloured wings and a similar body shape. This particular moth gets its name from the bright canary-yellow body that, with experience, allows separation from similar looking species. The single generation is on the wing July-October, favouring scrubby habitats, woodland, urban parks and gardens; the larvae feed on Hazel and other shrubs. It is common and widespread.

Lilac Beauty

SWALLOW-TAILED MOTH

Ourapteryx sambucaria
FOREWING LENGTH 22-30MM

With its characteristic pointed 'tails', this large yellow moth is unlike any of our other resident species. The 'tails' are pointed tips to the hind wings. Freshly emerged individuals are pale lemon yellow but fade to off-white with age. The single generation is on the wing June-August and can be found in a range of habitats, including urban gardens. The larvae feed on Ivy, hawthorns and other shrubs. Common across much of Britain, it extends north into the Scottish lowlands.

Silver-ground Carpet

Canary-shouldered Thorn

MOTHS, BUSH-CRICKETS AND BUGS

Many of our moths are well camouflaged and a careful search of bark and vegetation is needed to find them during the daytime. It is a similar story with some of the bush-crickets and bugs with which we share our gardens – despite the relatively large size and bright colours of some species they blend in well with their surroundings.

WINTER MOTH

Operophtera brumata
FOREWING LENGTH 13-19MM

As its names suggests, this moth is on the wing October–January and may be seen at lighted windows throughout the winter months. To be more precise, it is the males that are on the wing because the females are unable to fly, having tiny rudimentary wings. The larvae feed on a wide range of trees and shrubs and this may help explain why the moth is common and widespread.

MOTTLED UMBER

Erannis defoliaria
FOREWING LENGTH 18-25MM

The female of this common and widespread species is flightless; it is yellowish white in colour and can be found on tree trunks at night with the aid of a torch The male is fully winged and both sexes can be found October–January, the male coming readily to light. It is common in woodland and mature gardens. The larvae feed on broadleaved trees and shrubs.

MOTTLED BEAUTY

Alcis repandata
FOREWING LENGTH 19-26MM

This moth is variable both in terms of the ground colour of its wings and their patterning. The overall effect provides good camouflage when resting on tree bark in its favoured woodland and garden habitats. The Mottled Beauty is a common and widespread species throughout most of England, Wales and Scotland. It is on the wing June–August and the larvae feed on birches, oaks and Bramble.

PEPPERED MOTH

Biston betularia
FOREWING LENGTH 22-28MM

This moth is famous for its two colour forms. The white form, peppered with black, is found in rural areas, while the black form was once common in the highly polluted industrial cities. As air quality has improved, so the black form is being lost, its darker coloration no longer offering the camouflage it once did on soot-stained tree trunks. The adults are on the wing May–August and may be attracted to light. The larvae feed on various trees and shrubs.

Black form

White form

LEFT: **Mottled Beauty**
BELOW: **Oak Bush-cricket**

SPECKLED BUSH-CRICKET

Leptophyes punctatissima
LENGTH 9-18MM

This small, green bush-cricket can be identified by the minute dark spots that cover the body. A mainly southern species, commonest in southern England, the Speckled Bush-cricket favours scrub and rough vegetation. As such, it favours woodland rides, parks and mature gardens. Its speckled nymphs emerge in May and are often to be found sunning themselves on bramble leaves. They reach maturity in August. Adult females have a broad, sickle-shaped ovipositor, used for placing eggs in plant stems.

OAK BUSH-CRICKET

Meconema thalassinum
LENGTH 13-17MM

This small, pale green bush cricket can be found throughout the southern half of Britain, roughly north to the Dee and the Humber. It is our only completely arboreal bush-cricket, favouring oak and other broad-leaved trees. The Oak Bush-cricket may be encountered in gardens, occurring on shrubs like Hazel and Elder. On hot summer evenings, the species is often attracted to lights and ventures indoors through open windows.

DARK BUSH-CRICKET
Pholidoptera griseoaptera
LENGTH 13-20MM

The Dark Bush-cricket ranges in colour from a light brown to a very dark brown and has very short wings. In the southern part of its range, which just reaches into Scotland, it is often the most commonly encountered bush-cricket. The species favours scrubby habitats and can often be found, in some numbers, within bramble thickets. The nymphs hatch towards the end of April and reach maturity during July. The adults may then survive through into late November.

DOCK BUG
Coreus marginatus
LENGTH 12-14MM

This is a large brown bug, mottled in appearance and with a broad brown shield. Strictly speaking, it is not a shield bug but is very similar to them. Any habitat with lush vegetation and damper areas may be used and the species is common and widespread across southern England and Wales. It does, however, become rarer as you move further north. Adults may be encountered from August to the following July. Mating takes place April-July.

GREEN SHIELD BUG
Palomena prasina
LENGTH 12-14MM

This large, colourful shield bug is familiar to many gardeners: the body is mostly bright green with dark wingtips. Overwintering adults become brown but return to green the following spring. The Green Shield Bug is very common and widespread in central and southern England, becoming progressively scarcer further north. The larvae feed on broad-leaved trees June-October and adults are active from September to the following July.

BLACKFLY
Aphis fabae LENGTH 1-3MM

More correctly called the Black Bean Aphid, the 'blackfly' is widely regarded as a pest of garden plants. It passes the winter as an egg, from which aphid nymphs hatch February-April and colonise and form colonies on garden plants, feeding on young shoots and leaves. Later in the year, winged adult forms are produced and these disperse to find new feeding opportunities. A by-product of feeding colonies is the sticky secretion that may promote the growth of sooty mould.

COMMON FROGHOPPER
Philaenus spumarius
LENGTH 5-7MM

The Common Froghopper is the creature behind the watery collection of froth known as 'cuckoo spit' This froth is produced by the larva, which uses it as a form of defence. The adult froghopper is very variable in its coloration, ranging from almost black through to a straw brown, and has a mix of patterns. It may be found June-September on a wide range of plant species. Both the adult and the larva feed on the plant sap.

HAWTHORN SHIELD BUG
Acanthosoma haemorrhoidale
LENGTH 12-16MM

This is our largest shield bug, common and widespread across much of Britain, though scarce in Scotland. The red-brown tip to the abdomen and the 2 red projections at the shoulders should help with identification. The species hibernates as an adult and overwintered adults persist into June; the next generation of adults is not evident until late August. Various shrubs are used for feeding but Hawthorn is the favourite.

BRONZE SHIELD BUG
Eysacoris fabricii
LENGTH 5-6MM

Also known as the Woundwort Bug, on the basis of its use of Hedge Woundwort as a foodplant, this charming little bug is found across much of southern England and Wales. The species has been expanding its range over recent decades. The bug can often be found living in large colonies on a single foodplant, switching from Hedge Woundwort to Black Horehound on sandy soils and in urban areas. The larvae may be found from June-September.

ROSE APHID
Macrosiphum rosae LENGTH 1-3MM

As its name suggests, the Rose Aphid utilises domesticated roses in garden settings. It may also use teasel and scabious in the summer months. The aphids usually over-winter as eggs on rose stems but may, in the mildest winters, remain as active adults. Come spring, dense colonies develop, the foliage on which they establish becoming sticky with honeydew. The Rose Aphid is yellow-green or pink in appearance. Although regarded as a pest, Rose aphids are food for ladybirds, hoverflies and lacewings.

LACEWINGS, EARWIGS, FLIES AND BEES

Many of our most familiar garden insects – flies and bees for example – are winged and capable of sustained, active flight. Others, such as earwigs, are flightless and scuttle and climb through vegetation. Ant colonies comprise wingless individuals for much of the time, but winged forms appear in summer and aid dispersal.

LACEWING
Chrysoperla carnea
LENGTH 18-21MM

Of the 66 lacewing species found in Britain this is the one most likely to be found in gardens. And any green lacewing found indoors during the winter will be this species, since it is the only British lacewing to overwinter as an adult. Lacewings hold their wings up over their body like a tent, a feature shared with some true flies. Lacewing larvae are voracious predators and eat large numbers of aphids.

ABOVE: **Lacewing**

Lacewing larva

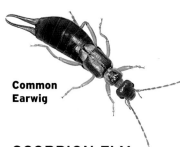

BELOW: **Scorpion Fly**

COMMON EARWIG
Forficula auricularia
LENGTH 10-15MM

The familiar earwig is an occasional pest in gardens but is otherwise harmless. The Common Earwig is widespread in Britain and it is commonly encountered in a range of habitats, including gardens. Female earwigs are devoted parents and lick each of their 30-50 eggs regularly to keep them clean and free from mould prior to hatching. Hatched young remain in the nest chamber with their mother for some time before becoming independent.

Common Earwig

SCORPION FLY
Panorpa communis
LENGTH 15-18MM

The male Scorpion Fly's swollen tipped abdomen gives this insect its name: it is often held above the body in a scorpion-like manner but is not a sting but the male's reproductive apparatus. The strongly marked wings and downward pointing 'beak' are also useful for identification, particularly in females, which lack the swollen abdomen tip. Scorpion Flies rest in shady locations on stands of brambles and nettles and are active May-August.

CRANEFLY
Tipula maxima
LENGTH 30-35MM

At 65mm, this handsome insect has a larger wingspan than any other British fly. Often referred to as 'daddy-long-legs', on account of the long and fragile legs, craneflies are an adaptable group of insects that can be found in most British habitats. While males have a blunt end to their abdomen, that of the female is pointed, helping her to lay eggs in the ground or some other material. The larvae of most cranefly species are soil dwellers and are often called 'leatherjackets'.

larva

DRONE FLY
Eristalis tenax
WING LENGTH 10-13MM

This abundant hoverfly derives its name from the fact that it mimics the drones (male hive bees) of the honeybee. Adults overwinter in sheltered crevices, outbuildings and (sometimes) houses. Because of this, they can be common in early spring but have been found in every month of the year. The larvae of these robust flies are aquatic, favouring organically rich drains and farm ditches; their long breathing tube earns them the name 'rat-tailed maggot'.

HOVERFLY
Helophilus pendulus
WING LENGTH 9-11MM

The thorax of this species has a series of black and yellow stripes running from front to back. Further black and yellow markings appearing on the abdomen aid identification but there are a number of other species that look very similar. This common hoverfly can be found at flowers in many sunny locations and, when perched, may emit a buzzing sound. The larvae occur in farmyard drains and in very wet manure.

HOVERFLY
Syrphus ribesii
WING LENGTH 7-12MM

This hoverfly also has yellow and black markings on the abdomen but the black thorax separates it from species of *Helophilus*. This particular species is one of the most common and widespread of our hoverflies, often encountered in gardens and on the wing from April to November. Males can often be found in dappled shade, defending a shaft of sunlight with a perceptible 'hum'. The larvae are often abundant at colonies of aphids, which they eat.

GREENBOTTLE

Lucilia caesar LENGTH 9MM

This is the most commonly encountered of a group of very similar looking flies. Very variable in size (which is dependent upon the amount of nutrients available to the growing larva), this species is rarely encountered indoors. The colour of this fly also varies (with age), starting out blue-green, the fly soon becomes emerald green before, near the end of its life, it becomes a dull coppery green. This species is most often seen on dead animals or excrement.

COMMON HOUSE-FLY

Musca domestica LENGTH 7MM

This fly has proved its adaptability by following Man all over the World. It breeds by using a range of moist household rubbish and, in many countries, human excrement. Each female will lay up to 900 eggs, which may hatch in as little as 8 hours, and the whole life cycle may be completed in less than 2 weeks. Although the species may be found throughout the year under suitable conditions, it is most commonly encountered from June through September.

RED-MASON BEE

Osmia rufa LENGTH 6-16MM

Widespread throughout much of Britain, the Red-mason Bee is a solitary bee and each nest is the work of a single female. The female builds her nest in an existing cavity, such as those made by beetles boring out of wood or abandoned lengths of garden cane. The species is active mid-March–July. Both sexes are covered with short ginger hairs; the females have a black face, the males have a white face.

GARDEN BLACK ANT

Lasius niger LENGTH 4-5MM

This familiar species forms colonies under paving and patio slabs in all types of garden, as well as in the countryside. Although called a 'black ant' it is really dark brown in colour. It is the ant species most likely to enter houses and forages widely away from the nest. Interestingly, it avoids bright sunshine, to the extent that it may build earthen covers to its more exposed trails. This species sometimes 'farms' aphids and collects their honeydew.

FLESH FLY

Sarcophaga carnaria LENGTH 14MM

This common and widespread fly requires carrion or other decaying animal matter for breeding. Unusually, the female deposits larvae rather than eggs into the decaying material that will sustain their development: she is viviparous, giving birth to live young. This species is grey with a pattern of dark markings on its body, which creates a chequerboard effect. It occurs in gardens but does not normally enter houses.

BLUEBOTTLE

Calliphora vomitoria LENGTH 11MM

The Bluebottle is a common and familiar fly, found in and around houses throughout the year. Within houses, the females seek out meat and fish for egg-laying but outside they use decaying flesh and excrement – one reason why their heavy buzzing is so detested by householders. They may sometimes lay their eggs on the wool of live sheep, close to a wound. Males sometimes visit flowers (especially umbellifers) for nectar.

LEAF-CUTTER BEE

Megachile centuncularis LENGTH 10MM

The most widespread of several related British leaf-cutter bees, this species is sometimes called the Patchwork Leaf-cutter Bee. It has bright brown hairs across the head and thorax, which then continue along the side of the abdomen. The characteristic disc-like leaf portions that are removed from roses and other plants are used by the bee in her nest building. The species' range is mostly southern and it is most noticeable June–July.

RED ANT

Myrmica rubra LENGTH 4-5MM

Of the 11 *Myrmica* species in Britain, this 'red ant' is one of the most widespread. It is most abundant in warmer southern counties and is noticeably less common as you move further north. *Myrmica rubra* forms small colonies, typically numbering fewer than 100 individuals. These are established under stones and in tree trunks and are often in damp habitats. This is one of a handful of ants that can deliver an unpleasant sting.

WASPS, BEES AND BEETLES

With their predatory habits, wasps fulfil an important role in the garden, controlling numbers of other insects. Bumblebees have more benign habits and are in decline generally – gardens an important refuge for them in conservation terms. Beetles, the most diverse of our insect groups, also have many garden representatives and are well worth studying.

HORNET

Vespa crabro
LENGTH 21-24MM (WORKERS)

Despite its size, this social wasp is surprisingly docile. It is widespread in southern Britain but rather uncommon and local within this range. The preferred habitat seems to be ancient woodland, with plenty of trees with holes in which the colony can nest. Nests may also be found, though less often, in roof spaces, bird boxes and empty bee hives. Hornets will feed on bees, but also ring bark shrubs and take nectar and pollen.

GERMAN WASP

Vespula germanica
LENGTH 12-15MM (WORKERS)

A familiar wasp species, often found in urban areas across Britain but absent from northwest Scotland. Seen head-on, the face is marked with 3 black dots. The nests are established in the rot holes of trees and in holes in the ground. The 'paper' that makes up the nests is invariably a uniform grey colour (that in Common Wasp is of mixed colour bands). By late summer the large nests may support a colony of several thousand individuals.

COMMON WASP

Vespula vulgaris
LENGTH 9-14MM (WORKERS)

As its name suggests, this is a common and widespread species. Seen head-on, the face is marked with a black anchor mark. Workers are attracted to sugary substances, which is why they may become a nuisance in those places where we eat food. Holes in the ground and trees are used for nesting, as are attics and garden sheds. The 'paper' used in the nest is produced from rotting wood and is a mixture of brown, cream and white bands.

MEDIAN WASP

Dolichovespula media
LENGTH 15-19MM (WORKERS)

The Median Wasp is a recent arrival to our shores, first recorded here in 1985. Since then it has spread across much of southern Britain. The face is marked with a dark median line down the centre. Nests are built suspended from the branches of a tree or shrub and are commonly reported from suburban gardens. When first established the nest has a small spout, whose function is unknown, but this disappears as the nest is increased in size.

Common Wasp nest

HONEY BEE

Apis mellifera LENGTH 10-14MM

Native to southern Asia, the Honey Bee is well-established in Britain, both in domesticated hives and in the wild. Wild colonies normally build their nest in a hollow tree but may, on occasion, build in the open. The colony is ruled by a queen who is only seen outside of the nest when the colony is swarming or on her mating flight. She is supported by a mass of workers and, in summer, males (drones) appear.

SAWFLIES

Sawflies derive their common name from the female's needle-shaped appendage used for egg-laying; in most species a saw-like action is employed and eggs are laid inside plant material. The larvae either feed openly on leaves or feed inside the plant itself, sometimes inducing the formation of characteristic galls. The adults of many species feed on pollen but some are partly carnivorous. They range in size from small species up to the magnificent Horntail, which can be 40mm in length.

Horntail

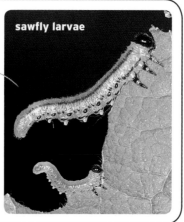

sawfly larvae

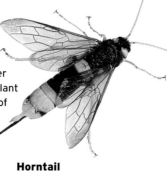

RED-TAILED BUMBLEBEE

Bombus lapidarius
LENGTH 22MM (QUEENS)

A recent and significant expansion has seen the range of this species increase within Scotland, adding to the wide distribution across England and Wales. Queens are large and black, with a red tip to the abdomen, and are on the wing from March onwards. The first workers are usually seen in May, the colony increasing in size to peak at several hundred individuals by August. Queens visit Dandelion flowers.

COMMON CARDER BUMBLEBEE

Bombus pascuorum
LENGTH 17MM (QUEENS)

The Common Carder Bumblebee is commonly encountered throughout Britain, favouring flowers with a deep flower tube. Queens are variable in colour and have long and untidy coats, ginger on the thorax, ginger and black on the abdomen. The bees emerge from hibernation a little later in the spring than other bumblebees, queens appearing in early April and the first workers produced during early May. The nest is often built using moss.

VIOLET GROUND BEETLE

Carabus problematicus
LENGTH 20-30MM

The Violet Ground Beetle is a large species, its metallic black body usually tinged with violet or sometimes blue. It can be found in many undisturbed habitats, including woodland and gardens, and occurs throughout Britain, only avoiding high ground. Unlike the larvae of many other ground beetles, those of the Violet Ground Beetle are active on the surface of the soil, where they appear to be at least partially predatory in their habits.

STAG BEETLE

Lucanus cervus LENGTH 25-75MM

This is our largest and most impressive beetle, the males carrying a set of imposing antler-like mandibles that are used in combat to secure access to females. The smaller females lack large mandibles. The larvae live in the root stocks of deciduous tree stumps and the species is mostly assocatied with oak woodland. The Stag Beetle has declined in Britain and it is now restricted to the south of England, where the adults may fly after dusk.

BUFF-TAILED BUMBLEBEE

Bombus terrestris
LENGTH 20MM (QUEENS)

The Buff-tailed Bumblebee is a common species, widespread throughout England and Wales. Its distribution within Scotland is more restricted, but seemingly increasing. The very large queens are black, with two yellow bands, one on the collar and one on the second abdominal segment. The tail is buff or off-white. Queens may be seen in early spring, searching for a nest site and moving low over the ground. The nest is usually underground in a bank or area of bare ground.

GROUND BEETLE

Pterostichus madidus
LENGTH 14-18MM

This shiny black ground beetle is found in woodlands, dry grasslands and gardens right across Britain and it is often extremely abundant. This species has a varied diet: it is mainly a scavenger but will also feed on rotting vegetable matter, ripe fruit and small invertebrates. Other ground beetle species are more active predators but some feed on seeds and plant material. Ground beetles are a varied group and identification of many species is best left to experts.

DEVIL'S COACH-HORSE

Ocypus olens LENGTH 22-32MM

The Devil's Coach-horse is a formidable beetle, one of the largest members of the rove beetle group (staphylinids). It is capable of delivering a powerful and painful bite and when threatened it opens its jaws and curls its tail end, an intimidating ploy that usually deters further disturbance. It favours woodland, grassy places and gardens, and is usually found under stones and logs since it is nocturnal in behaviour. Slugs and other invertebrates are the main prey.

WASP BEETLE

Clytus arietus LENGTH 7-14MM

The characteristic yellow and black pattern on the wing cases of this longhorn beetle is what gives this species its name. This pattern, combined with the way in which the beetle moves its legs and antennae, gives the impression of a wasp; presumably this form of mimicry is used by the beetle to deter would-be predators. Adult Wasp Beetles may be seen visiting flowers May–July; their larvae spend up to 2 years living within dead wood prior to emergence.

Beetles are an extremely diverse group of insects and it is little wonder that many representatives are found in the garden. Depending on the species, their diet may be carnivorous or herbivorous but spiders, by contrast, are all active predators and, for smaller invertebrates, a force to be reckoned with in the garden.

ROSE CHAFER

Cetonia aurata LENGTH 14-20MM

This striking beetle is a southern species, local in occurrence across the south of England then becoming increasingly rare as you move further north. Adults overwinter in soil rich in decomposing matter and may be seen in spring from April onwards, often flying in the afternoon and then settling on flowers later in the day. The species is declining.

COCKCHAFER

Melolontha melolontha
LENGTH 20-30MM

Popularly known as the 'may-bug' this species is scarcer than it used to be. It is found across southern Britain and adults may be more abundant in some years than others. The larvae feed on roots and may sometimes be a major pest of amenity turf or cereal crops. The adults feed on the leaves of trees. The life cycle lasts about 4 years; most time is spent as a larva.

LILY BEETLE

Lilioceris lilii LENGTH 6-8MM

This fairly common species has a distinctive colour and shape, and may be found on its host plants April-July. Although there are records from Scotland, the main range is southeast England. It may be found in gardens and commercial nurseries, occurring on lilies and fritillaries, where it is regarded as a serious pest. While the adults damage the leaves of food plants, the larvae can decimate leaves and buds.

MINT LEAF BEETLE

Chrysolina herbacea
LENGTH 8-10MM

This beetle, which is usually a bright metallic green in colour, occurs in damp habitats and gardens, where it feeds on a range of mint species, especially Water Mint. It is mainly encountered in central southern England but there are records from across Britain, possibly the result of horticultural trade. The adults first emerge in late June and can overwinter successfully, often living for several years. Most records refer to adults, seen June-September.

larva

VINE WEEVIL

Otiorhynchus sulcatus
LENGTH 8-10MM

The Vine Weevil is a major horticultural pest, feeding on a very wide range of plants. The larvae are root feeders and can do significant amounts of damage to plants, especially those being grown in pots, while the adults damage leaves. A number of closely related species have now been accidentally imported into Britain but the Vine Weevil is the most widely distributed of these. It is found across England and Wales, its range extending into Scotland.

EYED LADYBIRD

Anatis ocellata LENGTH 8-9MM

The Eyed Ladybird is one of our largest ladybird species but is often overlooked because it has a particular association with pine trees. However, it may be encountered in those gardens with or near to mature pines. The red or orange-red wing cases usually carry 18 spots, each finely edged with a narrow ring of white or pale yellow. This ladybird has 1 generation per year, with fresh adults most often encountered towards the end of July.

7-SPOT LADYBIRD

Coccinella 7-punctata
LENGTH 5-8MM

The 7-spot Ladybird is our most familiar species and the one so often portrayed in children's literature. It is a common and widespread species, occurring across most of Britain. Interestingly, while the adult is well known and readily found, its larva largely goes unrecognised: grey in colour, with eight pale spots, it has an elongated abdomen and biting mouthparts. It can often be found close to the aphid colonies on which it feeds.

larva

2-SPOT LADYBIRD

Adalia bipunctata LENGTH 4-6MM

This species seems to have a stronger association with urban habitats than our other ladybirds and can often be found in suburban gardens. Like the 7-spot, it is a familiar predator of aphids and is very much the gardener's friend. The typical adult form has a single black spot on each red wing case, but there are melanic forms, with 4 or even 6 red spots on a black background. Ladybirds generally are well known for such variation.

2-spot Ladybird dark form

14-SPOT LADYBIRD
Propylea 14-punctata
LENGTH 5-6MM

The 14-spot Ladybird has a series of black spots on a creamy-yellow background. The spots themselves are often fairly square in appearance and typically join up to form a distinct pattern, reminiscent of a clown's face or an anchor. Adults are normally seen April–September, with a dip in abundance in July that suggests 2 distinct generations each year. It can be found on tall herbaceous vegetation and is common across most of the country.

HARLEQUIN LADYBIRD
Harmonia axyridis LENGTH 5-8MM

A recent and unwelcome addition to our ladybird fauna, this species arrived with imported cut flowers from continental Europe. It is a large species, highly variable in colour and with the number of spots varying from 0 to 21. This species is a problem because it does not just feed on aphids; when aphid numbers are low it turn its attentions to other insects including ladybirds. Its presence, therefore, may threaten various native species.

GARDEN SPIDER
Araneus diadematus LENGTH 10-18MM (FEMALE), 4-8MM (MALE)

This common and widespread species may be found on vegetation throughout summer and autumn, in habitats ranging from woodland to gardens. Although there is considerable variation in the ground colour (pale cream to almost black) the distinctive abdomen markings are usually visible and aid identification. Like other orb-web spiders, it spins orb-shaped webs that catch flying prey. Females may eat their suitors.

CRAB SPIDER
Misumena vatia LENGTH 9-11MM (FEMALE), 3-4MM (MALE)

This distinctive spider sits in flowers and ambushes visiting insects. The colour of the spider, typically white, yellow or green, allows it to blend in with the flower petals on which it is positioned, hence it is commonly be found on flowers that are pale yellow or white. The spider is able to change colour, albeit slowly, further adding to its deception. Found across England and Wales, it is commonest in the south.

ZEBRA SPIDER
Salticus scenicus LENGTH 5-7MM

This small spider is the most commonly encountered of our three 'zebra' spiders, occurring across Britain and much of Europe. Within gardens, it is usually to be found on walls or fences, where its criss-cross draglines may cover the surface. The spider itself tends to hide away in a suitable crevice, only becoming active when the sun comes out and the substrate warms up. An alert spider, it will watch your approach by raising itself up to get a better view.

HOUSE SPIDER
Tegeneria domestica LENGTH 9-10MM (FEMALE), 6-9MM (MALE)

One of several similar species, this spider is cosmopolitan and may be found in houses across Britain. The sheet webs produced in undisturbed corners of the house can reach a considerable size and this is where the female spends much of her time. Individuals seen dashing across the carpet in late summer are usually males wandering in search of a mate. They will remain with her for some weeks, before they die of old age and get devoured.

HARVESTMEN

Despite a superficial resemblance to spiders, harvestmen belong to a completely different group of invertebrates. Unlike spiders, they do not produce silk, nor do they produce venom. The adults mature in late summer, hence the name harvestman, and this is when they are most often encountered. Our 20 or so different species are largely nocturnal and may be found in a range of different habitats including gardens. They feed on a wide range of other organisms and may also scavenge dead insects.

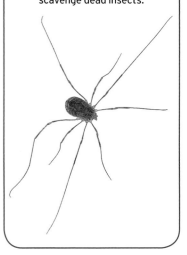

FRUIT TREE RED SPIDER MITE
Panonychus ulmi

The Red Spider Mite belongs to a group of arachnids that also includes ticks. They are characterised by their typically very small size and short legs. It is a serious pest of fruit trees and may often be encountered in gardens across much of Britain. The first eggs, placed on the food plant, hatch in April and there may be 5 or more other generations throughout the rest of the year. A different species is a known pest in glasshouses.

MOLLUSCS, WOODLICE, CENTIPEDES & MILLIPEDES

Slugs and snails are present in almost every garden in Britain and, although reviled by many gardeners, only a few species cause problems. Like woodlice, centipedes and millipedes, most lead rather unobtrusive lives, becoming active mainly after dark.

LARGE RED SLUG
Arion ater LENGTH 100-150MM

While the size of this slug may match its name, its colour often doesn't. It occurs as several distinct colour forms, ranging from orange, through brick red to grey and jet black; the black form is commonest in the north of its range. This slug is catholic in its choice of habitats and may be found even on the most acid ground. The foot fringe is paler in colour than the body and is often red in appearance.

LEOPARD SLUG
Limax maximus
LENGTH 100-200MM

Widespread across Britain, the Leopard Slug is a very large slug, usually pale brown or grey in colour, with several darker bands on each side and often a marbled pattern on the mantle. The tentacles are a uniform reddish brown. A range of habitats, including gardens and woodland may be used and the species sometimes even ventures indoors, squeezing through ill-fitting doors.

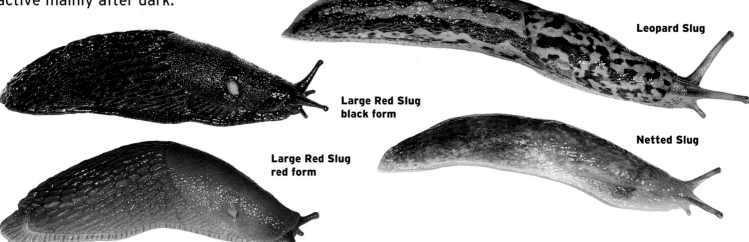

Leopard Slug

Large Red Slug black form

Netted Slug

Large Red Slug red form

YELLOW SLUG
Limax flavus LENGTH 75-100MM

The Yellow Slug shows a strong association with Man and is often found in kitchens, damp cellars, outbuildings and gardens. Even though it is widespread across much of Britain, this species is local. The body is usually pale yellow overlaid with grey markings. While slime produced by the foot is colourless, that produced by the body is yellow. Nocturnal in habits, it tends to emerge when conditions are most favourable, usually when the air is damp following rain.

NETTED SLUG
Deroceras reticulatum
LENGTH 35-50MM

This medium-sized slug is probably the most abundant slug across much of lowland Britain, and is a major pest of agriculture. The species is equally at home in grassland, along hedgerows and in gardens. The typical form is pale cream or pale brown with a dense pattern of darker speckles; some pale forms lack these speckles, leading to confusion with other species. The slug is well distributed across Britain.

GARDEN SNAIL
Helix aspersa
SHELL HEIGHT 25-35MM

This is the most familiar of our snail species and is found in a range of habitats, including gardens. Here it may sometimes be regarded as a pest. The shell is usually pale brown, with up to five dark spiral bands. These are variable in colour and sometimes flecked with white. The species is found across Britain, though it is restricted in Scotland to the lowlands and east coast fringe.

STRAWBERRY SNAIL
Trichia striolata
SHELL HEIGHT 7-9MM

The Strawberry Snail is a common species in gardens across much of Britain, though most numerous in the southern part of its range. In addition to gardens, it uses waste ground, woodlands and hedgerows, favouring lowlands and avoiding particularly dry habitats. The opaque or slightly translucent shell is somewhat flattened, with rough growth ridges.

Garden Snail

Strawberry Snail

Yellow Slug

COMMON SHINY WOODLOUSE

Oniscus asellus LENGTH 16MM

This species is normally a shiny grey colour, with irregular light markings and 2 rows of pale yellow patches. However, orange and yellow forms have been recorded in coastal areas. It may be encountered in a range of damp habitats, notably by compost heaps and in other garden refuse. The Common Shiny Woodlouse is extremely common throughout Britain and Ireland and may often be found in the company of the Common Rough Woodlouse.

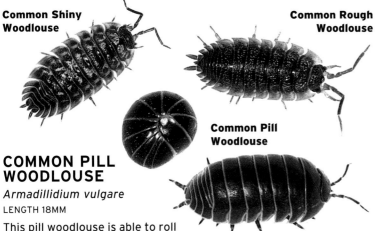

Common Shiny Woodlouse

Common Rough Woodlouse

Common Pill Woodlouse

COMMON PILL WOODLOUSE

Armadillidium vulgare LENGTH 18MM

This pill woodlouse is able to roll itself into a completely enclosed ball, and is the only species to be able to do this; 2 related species both leave a gap when they roll into a ball. It is extremely common in the south and east of England but is more coastal in its distribution elsewhere. It is absent from much of northwest England and Wales and Scotland. The species is usually a uniform slate grey in colour.

COMMON PYGMY WOODLOUSE

Trichoniscus pusillus LENGTH 5MM

This species is probably the most abundant woodlouse in lowland woodland, favouring damp soil and leaf litter. It is purple-brown or reddish brown in colour but individuals with a rich purple sheen are sometimes seen, the colour the result of a viral infection. Intriguingly, it occurs in 2 forms, one where males and females occur in equal numbers, the other comprising mostly females.

COMMON ROUGH WOODLOUSE

Porcellio scaber LENGTH 17MM

Burrowing Centipede

The Common Rough Woodlouse is very common throughout much of its wide British range, occupying a variety of damp habitats. More tolerant of dry conditions than many other woodlice, it may sometimes enter houses. In colour it is usually slate-grey, but cream and orange forms may be found near to the coast. When first disturbed it tends to remain motionless but will then run away with a surprising turn of speed.

CENTIPEDE

Lithobius forficatus LENGTH 20MM

Like other centipedes, this species is an active surface predator and uses its venomous bite to kill suitable prey – smaller invertebrates. In the daytime it is often found hiding under logs and stones, and behind flaking bark, and it is common in gardens throughout the region. A similar species, *L. variegatus*, a mainly woodland species, sometimes turns up in rural gardens, particularly in western Britain.

Common Pygmy Woodlouse

BURROWING CENTIPEDE

Stigmatogaster subterranean LENGTH 150MM

This long, worm-like centipede lives a subterranean existence, moving through the spaces between soil particles in search of prey. It is often discovered when digging the vegetable patch. Living underground, it has no need for eyes and is effectively blind. The Burrowing Centipede is widespread in Britain and is virtually ubiquitous in gardens across the region.

COMMON EARTHWORM

Lumbricus terrestris LENGTH 8CM

One of many species of earthworm found in the region and often common in the garden. Recognised by its elongated, segmented body, tapering at one end, rather blunt-ended at the other. Essential for soil aeration and fertility (leaves and organic debris is removed from the soil surface and buried). Important as food for birds such as Blackbird and Song Thrush. Conspicuous deposits of digested soil casts are deposited at burrow entrances.

BLUNT-TAILED SNAKE MILLIPEDE

Cylindroiulus punctatus LENGTH 25MM

This slow moving millipede is sometimes encountered in rural gardens although woodlands are its favoured habitat. It is known to utilise both soil and rotting wood, moving between the two for different stages of its life cycle. Like other millipedes, this species may roll itself into a coil when threatened. The pale brown body is darker above and has a series of dark markings along the sides – 1 per segment.

Blunt-tailed Snake Millipede

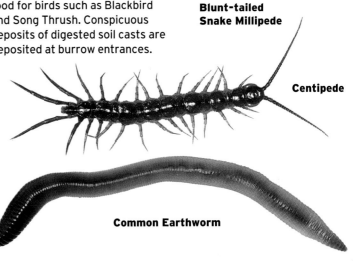

Centipede

Common Earthworm

SPARROWHAWK
Accipiter nisus

adult male

juvenile

Male is colourful with blue-grey upperparts, orange-flushed face and reddish barring on underparts; staring orange-yellow eyes are striking.

In flight, note broad, rounded wings, relatively long tail and barred underparts.

PHEASANT
Phasianus colchicus

male

Male is unmistakable with mostly orange-brown plumage, long tail and greenish sheen on head and neck; red facial wattles are striking.

male and female

In spring, males remain in close contact with females, keeping a watchful eye open for rivals.

Drab, brownish plumage affords female a degree of camouflage when nesting.

female

BLACK-HEADED GULL
Larus ridibundus

summer adult

In winter, head is mostly white with dark smudges behind eye; leg and bill colours are often paler than in summer.

Dark-tipped tail and patterned upperwings allow separation from winter adult.

Chocolate-brown hood is diagnostic in summer months; legs and bill are deep red.

winter adult

1st winter

FERAL PIGEON
Columba livia

adult

adult

Dark wingbars contrast with otherwise pale grey upperwings; violet and green sheen seen on neck in good light.

Broad tail and wings allows for manoeuvrability in flight.

Adult has grey-brown upperwings and back contrasting subtly with pink-flushed underparts; green and white markings on greyish neck are striking.

adult

White wing patches are particularly striking in flight.

WOODPIGEON
Columba palumbus

adults

adult

COLLARED DOVE
Streptopelia decaocto

Plumage grades from pinkish buff on head, neck and underparts to buffish brown on back and upperwings; black and white neck marking is striking.

adult

White margin to tail is striking when fanned in flight and used in display.

TAWNY OWL
Strix aluco

Rich brown plumage is finely marked with dark bars and stripes; facial disc is round and eyes are dark.

adult

adult

Broad, rounded wings and special feather margins allow for silent, well-controlled flight.

adult

LITTLE OWL
Athene noctua

Flight is usually low and gliding, on broad, rounded wings.

adult

Upright posture and rounded, large-headed profile are clues to identification of perched birds; plumage is subtly marked with pale spots and eyes have bright yellow iris.

SWIFT
Apus apus

adult

Cigar-shaped body is obvious when seen in level flight.

adult

Wings held in swept-back, horseshoe shape when gliding fast.

KINGFISHER
Alcedo atthis

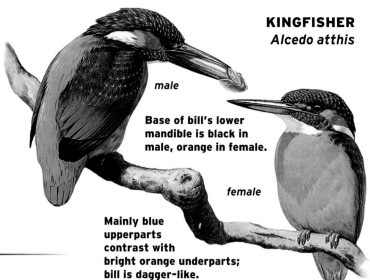

male

Base of bill's lower mandible is black in male, orange in female.

female

Mainly blue upperparts contrast with bright orange underparts; bill is dagger-like.

female

RING-NECKED PARAKEET
Psittacula krameri

adult male

Plumage is mostly bright green, grading to bluish on the uppertail; only the male has a neck 'ring'.

adult male

Dark underwings and long tail are obvious in flight.

GREEN WOODPECKER
Picus viridis

female

Yellow rump sometimes seen in climbing birds; female has dark malar stripe.

male

Yellow rump is striking in flight.

male

Has green upperparts, red cap and mainly black face; male has red centre to black malar stripe.

male

GREAT SPOTTED WOODPECKER
Dendrocopus major

Mainly black and white plumage with red on nape and undertail.

White barring on wings and bold white 'shoulders' seen in flight.

male

Similar to male but lacks red on nape.

female

SWALLOW
Hirundo rustica

adult

Juveniles have paler throats and shorter tail streamers than adults.

begging juvenile

adult

Upperparts have blue sheen; forked tail and long streamers are obvious in flight.

adult

HOUSE MARTIN
Delichon urbica

Upperparts are blue-black, except for white rump, and underparts are whitish; tail is forked.

adult

White rump contrasts strikingly with blue-black upperparts.

PIED WAGTAIL
Motacilla alba

Black, grey and white plumage and long tail are diagnostic; male has more black on head than female and more in summer than winter.

male

male

Flight is bounding and often accompanied by characteristic 'chissik' call.

WAXWING
Bombycilla garrulus

Black facial markings, chestnut undertail and red 'wax' tips to inner wing feathers are obvious in adults.

adult

Plumage grades from pinkish buff to lilac; peaked crest is diagnostic. Male has more extensive yellow and white markings on wing than female.

adult male

WREN
Troglodytes troglodytes

adult

Has rounded body and barred brown plumage; tail is often cocked up.

adult female

ROBIN
Erithacus rubecula

Orange-red face and breast is diagnostic; upperparts are otherwise brown, underparts whitish grading to pale grey.

adult

juvenile

Lacks adult's red breast; upperparts adorned with pale spots.

BLACKBIRD
Turdus merula

female

Plumage is overall reddish brown with faint streaks and spots on the throat and underparts.

male

Jet-black plumage and yellow bill and eyering.

adult

Blue-grey head, chestnut back and yellow-flushed spotted underparts are distinctive.

FIELDFARE
Turdus pilaris

adult

In flight, pale rump contrasts with chestnut back and dark tail.

adult

Upperparts are mainly brown, with bold white supercilium; spotted underparts are flushed red on flanks.

REDWING
Turdus iliacus

Red on flanks extends to underwing coverts and is obvious in flight.

adult

SONG THRUSH
Turdus philomelos

Orange-buff underwing coverts are obvious in flight.

adult

Upperparts are warm brown with two buff wingbars; underparts are overall pale with dark spots, flushed orange-buff on throat and flanks.

adult

MISTLE THRUSH
Turdus viscivorus

adult

Plump-bodied bird with brown upperparts and spotted pale underparts, flushed orange-buff on breast and flanks.

White underwings coverts and white tips to outer tail feathers are distinctive in flight.

adults

DUNNOCK
Prunella modularis

adult

adult

Often sings from exposed perch in spring, when streaked brown back and bluish head are obvious.

Forages unobtrusively, often adopting rather horizontal stance.

WHITETHROAT
Sylvia communis

FAR LEFT: *adult male*

Chestnut back, white throat and greyish head are striking; female is duller than male.

LEFT: *juvenile*

Similar to adult, with white throat and chestnut on back; head is brownish, most similar to adult female.

BLACKCAP
Sylvia atricapilla

male

Combination of black cap, greyish neck and underparts, and brown back are diagnostic among warblers.

Chestnut cap is common to both adult females and immatures of both sexes.

female

CHIFFCHAFF
Phylloscopus collybita

adult

Bill is narrow and needle-like.

adult

Plumage is rather undistinguished, grey-brown overall and darker above than below.

GOLDCREST
Regulus regulus

FAR LEFT: *male*

Body is nearly spherical with greenish upperparts, paler underparts and striking wingbars; yellow crown stripe is reddish-centred in male.

LEFT: *female*

Similar to male but crown stripe is uniformly yellow.

SPOTTED FLYCATCHER
Muscicapa striata

FAR LEFT: *adult*

Plumage is undistinguished grey-brown with two pale wingbars and streaked crown; often adopts a rather upright posture when perched.

LEFT: *juvenile*

Superficially similar to adult but underparts and back are subtly more spotted.

BLUE TIT
Cyanistes caeruleus

FAR LEFT: *adult*

Green and blue upperparts, mostly yellow underparts and white face make for easy recognition.

LEFT: *juvenile*

Resembles adult but plumage is overall grubby and face is yellowish.

GREAT TIT
Parus major

FAR LEFT: *adult*

Greenish back and black and white on head; underparts are yellow with black stripe down centre of breast, broader in male than female.

LEFT: *juvenile*

Plumage overall grubbier than adult, with yellowish face.

MARSH TIT
Poecile palustris

adult

Rather small black 'bib'.

Has glossy black crown, white face, grey-buff upperparts and buffish-white underparts.

adult

WILLOW TIT
Poecile montanus

adult

Black 'bib' is usually larger than in Marsh Tit.

adult

Pale wing panel seen in most birds; dull crown and thick neck are pointers for separation from Marsh Tit.

COAL TIT
Periparus ater

adult

White cheeks are striking when viewed from the front.

Warbler-like appearance with narrow bill, grey upperparts with white wingbars, and buff-flushed underparts; blackish cap has white central stripe.

adult

LONG-TAILED TIT
Aegithalos caudatus

adult

Has fluffy-looking, almost spherical body, tiny bill and very long tail.

adult

Long tail and rounded wings are obvious in flight.

NUTHATCH
Sitta europaea

adult

The only British bird that routinely climbs down tree trunks head first.

Body is rather rounded with a pointed, chisel-like bill; upperparts are blue-grey, underparts reddish buff.

adult

Streaked and speckled brown plumage affords good camouflage against tree bark; long, downcurved bill used to probe for insects.

TREECREEPER
Certhia familiaris

JAY
Garrulus glandarius

adult

Plumage is overall pinkish buff, black wings having striking blue and white markings.

White rump is diagnostic in a bird of this size.

adult

MAGPIE
Pica pica

In flight, wings are rounded with striking black-and-white markings; tail is long and tapering.

adult

Black-and-white plumage and long tail make recognition easy; blue and green sheen seen on wings and tail.

adult

JACKDAW
Corvus monedula

Can look all dark in poor light but if seen well grey nape and underparts are noted; eye has pale iris.

adult

adult

Aerobatic in flight, wings having 'fingered' tips (primaries).

CARRION CROW
Corvus corone

adult

adult

Plumage is overall black, with sheen on wings and back seen in good light.

Wings are broad and long, with 'fingered' tips (primaries).

HOODED CROW
Corvus cornix

Grey on back and belly contrast with otherwise black plumage.

adult

adult

Contrast between grey and black elements of plumage is striking in flight.

ROOK
Corvus frugilegus

adult

adult

adult

Wings and tail look proportionately long in flight; wingtips are 'fingered' (primaries).

Glossy black plumage has a sheen in good light; bill is long with pale, bare skin at base.

Summer bird has fewer spots than in winter and bill is yellow.

winter adult

Bill is dark and glossy plumage is adorned with numerous pale spots.

winter adult

Wings look triangular in outline in flight.

summer male

STARLING
Sturnus vulgaris

FAR LEFT: *male*

Has reddish brown streaked back and grey-buff underparts; intensity and extent of black on head varies with season and age.

LEFT: *female*

Upperparts streaked grey-brown and underparts grey-buff; head lacks male's black markings.

male

White wingbars contrast with reddish brown on upperwings in flight.

HOUSE SPARROW
Passer domesticus

Reddish-chestnut back and crown, and pale wingbars, are obvious in flight.

adult

Combination of chestnut crown, black ear patch on otherwise white face, and black bib are diagnostic; back is streaked reddish chestnut.

adult

In flight, the white cheeks with their central black ear patch are surprisingly easy to spot.

TREE SPARROW
Passer montanus

male

Plumage is overall greenish, with striking yellow line seen on wings at rest.

female

Overall less colourful than male, particularly pale on underparts.

GREENFINCH
Carduelis chloris

Black tip and yellow sides to tail are striking in flight.

adults

GOLDFINCH
Carduelis carduelis

adult

Combination of red, white and black on face, and yellow flashes on wings, are diagnostic; upperparts are otherwise buff, underparts pale.

juvenile

Yellow on wings is obvious but lacks adult's distinctive face markings.

adult

Yellow wingbar is striking in flight.

SISKIN
Carduelis spinus

RIGHT: *male*

Colourful finch with streaked yellow plumage overall, palest below, and striking wingbars; dark cap varies in intensity according to season.

FAR RIGHT: *female*

Colourful, with streaked yellowish plumage; lacks male's black markings on head.

Buffish-brown plumage overall, darker above than below, with pale wingbars.

CHAFFINCH
Fringilla coelebs

Pale wingbars and greenish rump are obvious in flight.

male

female

male

Colourful plumage with chestnut on back, blue on crown and nape, and pinkish-orange underparts; wingbars are striking.

BRAMBLING
Fringilla montifringilla

FAR LEFT: *winter male*

Intensity of black hood and back increase (through wear of pale tips) through winter; precise shade of orange-buff on breast and 'shoulders' is diagnostic.

LEFT: *winter female*

Colour is less intense than on male and head and back are buffish grey not black.

winter male

White rump is a good identification feature in flight.

male

Bluish-grey back, rose-pink underparts and black cap make identification straightforward; bill is stubby.

female

Similar to male but with subdued colours.

BULLFINCH
Pyrrula pyrrhula

male

White rump and contrasting black tail are striking in flight.

LESSER REDPOLL
Carduelis cabaret

male

Has red forecrown, pinkish flush to face and throat, and streaked brown upperparts with pale wingbars; underparts are heavily streaked.

COMMON REDPOLL
Carduelis flammea

male

Buffish-white plumage is much paler overall than Lesser Redpoll.

Variable but usually paler than male, with less intense red on head.

Lesser Redpoll female

Lesser Redpoll female

Rather indistinct wingbars and forked tail are noted in flight, which is often accompanied by rattling call.

REED BUNTING
Emberiza schoeniculus

summer male

Has streaked reddish-brown back and wings, and streaked pale underparts; black hood with white 'moustache' is seen in summer but lost in winter.

female

Female's plumage and colour recalls male's; head pattern is striking. In flight, upperparts appear uniform brown and streaked.

female

BIRDFACTS

FAMILY	Accipitridae
LENGTH	33cm
WINGSPAN	62cm
WEIGHT	Male 150g; female 260g
HABITAT	Woodland, farmland and gardens.
FOOD	Small birds.
NEST	Loose structure of twigs, with deep cup, built in tree canopy.
EGGS	4-6 eggs; rounded oval, bluish white and streaked reddish brown.
EGG SIZE	40 × 32mm
INCUBATION	33 days. 1 brood.
BREEDING SEASON	April to August
POPULATION SIZE	39,000 pairs.
TYPICAL LIFESPAN	3 years
MAX. RECORDED AGE	17 years 1 month

SPARROWHAWK
Accipiter nisus

With its slender body, short broad wings and long tail, the Sparrowhawk is ideally adapted to hunting within woodland. Now that its population has recovered from decades of persecution, the Sparrowhawk is a species that is seen increasingly in gardens.

CHANGING TIMES

The Sparrowhawk is now one of the most abundant and widely distributed birds of prey in Britain but just a few decades ago it was all but extinct in many eastern counties, following the introduction of organochlorine pesticides for use in agriculture. Such compounds are incredibly persistent and accumulate in the bodies of predators like the Sparrowhawk. One of these compounds, DDT, brought about a thinning of Sparrowhawk eggshells; in turn, this led to increased breakages during incubation and a decline in breeding success. The consequence was a massive drop in the breeding population, which only began to recover once these chemicals had been phased out. Even today, the legacy of DDT is still detectable in Sparrowhawk eggs. The current Sparrowhawk population is probably larger than it was before the decline, partly because of reduced levels of persecution but also a consequence of the establishment of conifer plantations in upland areas, which the birds now occupy. Recent decades have also seen a spread into many urban and suburban parks and gardens, with Sparrowhawks exploiting the abundance of small birds using garden feeding stations.

The long legs and extended central toes are adaptations that enable the Sparrowhawk to catch other birds.

A CLEVER HUNTER

Cover is important for a hunting Sparrowhawk and an individual will often work a hedgerow in order to get as close as possible to potential prey. Species that are conspicuous or which feed out in the open are most often taken. By moving your feeding station around your garden, and by keeping it close to cover, you can reduce the chances of a Sparrowhawk catching your birds by surprise.

Perched Sparrowhawks, such as this male, often adopt an upright posture.

BATHING

Regular bathing is a means by which birds can keep their all-important feathers in top condition. While some birds use dust or dew for bathing, many others, including Sparrowhawk, prefer to bathe in water, favouring shallow pools and bird baths.

THE BUSINESS OF BREEDING

Woodland is the favoured breeding habitat for Sparrowhawks and many pairs choose to nest in fairly dense woodland. Egg-laying begins several weeks after the nest has been built and research has shown that it usually starts 5-10 days after the first fledgling songbirds appear in the Sparrowhawk's diet. This dependency upon the presence of fledgling songbirds to initiate breeding highlights the close association between predator and prey. Food shortage is the commonest cause of Sparrowhawk nest desertion in Britain. During incubation, and while the chicks are still young, the female remains on the nest, supported by her mate who brings food for her and the developing chicks. Once the young are larger, both parents will bring food back to the nest, the female only brooding the chicks when the weather is poor. Young chicks grow quickly and can leave the nest after four weeks, although their parents support them for another three to four weeks after this.

A female Sparrowhawk at the nest.

IT'S A FACT...

The colour of a Sparrowhawk's eye varies with gender and may deepen with age. Typically, young birds have greenish-yellow eyes, which become a brighter yellow within the first couple of years of life. In some older males, the eye colour deepens further to orange or, occasionally, wine red.

RED IN TOOTH AND CLAW

The Sparrowhawk is a highly efficient hunter that specialises in catching other birds. Because of this, the Sparrowhawk is sometimes regarded as a villain - the perceived reason for the widespread decline of many songbirds. However, all the scientific evidence gathered so far strongly suggests that the Sparrowhawk is not to blame for the declines seen in other species. Other factors, notably changes in the nature of our farmland and woodland habitats, are more likely to be the driving force behind the declines.

BIG IS BEAUTIFUL

At 260g, a female Sparrowhawk is about twice the weight of her mate. Although it is not uncommon for female raptors to be larger than males, the Sparrowhawk shows the largest such difference of any bird of prey. The larger female carries the extra body reserves needed for reproduction and she can survive for a number of days without a meal. However, she is a less agile hunter than the male, whose role is to provision his mate. The two sexes focus on differently sized prey; males take smaller birds like finches and tits, only occasionally taking birds as big as a Blackbird. Females tend to take Starlings and larger birds up to the size of an adult Woodpigeon.

Female Sparrowhawk eating Woodpigeon prey.

A Sparrowhawk's broad, rounded wings and relatively long tail make flight identification straightforward.

A mature male Sparrowhawk has richly colourful plumage.

STATUS

Rapid increase in population since 1970s, though growth now levelled off. Green listed.

SEASONAL TRENDS

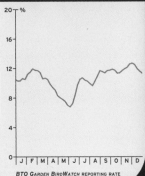

BTO GARDEN BIRDWATCH REPORTING RATE (%) THROUGHOUT THE MONTHS OF THE YEAR

PHEASANT
Phasianus colchicus

FAMILY
Phasianidae

LENGTH
71cm

WINGSPAN
80cm

WEIGHT
1.4kg

HABITAT
Farmland, woodland and scrub.

FOOD
Omnivorous, taking seeds, berries, leaves, roots and small invertebrates from the ground.

NEST
A slight depression in the ground, usually unlined but hidden in thick cover.

EGGS
5–12 creamy white eggs, unmarked and broad oval in shape.

EGG SIZE
46 × 36mm

INCUBATION
23–28 days. 1 brood.

BREEDING SEASON
March to August

POPULATION SIZE
1.7 million wild-breeding females, though up to 25 million birds may be released each year.

TYPICAL LIFESPAN
No data.

MAX. RECORDED AGE
2 years 1 month

The Pheasant is the commonest gamebird in Britain and the one most likely to be encountered within a garden environment, especially in late winter. Small groups may visit garden feeding stations to take seeds and grain, where the males add a welcome splash of colour.

FOODS AND FEEDING
Garden feeding stations may provide visiting Pheasants with seeds and grain. These foods, together with berries, plant tubers and weed seeds, are particularly important throughout the winter months. With the arrival of spring, Pheasants begin to increase the amount of animal food in their diet, taking a range of insects and other invertebrates. Although they may be seen feeding on bird tables, Pheasants prefer to feed on the ground, scratching or digging for food. They may sometimes jump to reach berries hanging just out of reach, a behaviour that may explain why BTO Garden BirdWatchers sometimes report Pheasants jumping up at hanging seed feeders and knocking seed free, which can then be gathered from the ground below.

> **IT'S A FACT...**
> Half of all gamebirds shot each year are pheasants, some 12 million birds. As such, shooting is a multi-million pound industry that has helped to shape our countryside.

FOREST ORIGINS
Originally birds of dense forest, Pheasants were probably introduced to Britain by the Normans; they were certainly present across much of England and Scotland by the late 16th century. In Wales, the species only really established itself fully during the latter part of the 19th century, while in Ireland it remains relatively uncommon and probably only survives through reinforcement. The greatest numbers of Pheasants are to be found on and around large shooting estates, where management often involves the hand rearing and release of thousands of birds annually. It has been estimated that up to 25 million Pheasants may be released for shooting each year, a figure that dwarfs the 'wild' population. Over the years, a number of different forms have been introduced to Britain in the hope that they might make better sporting birds. These forms have interbred with each other and so there is quite a bit of variation in the plumage of the familiar ring-necked males. One particular variety, popular over recent years, is the 'blue-back'; these are slightly smaller than normal birds and therefore fly higher and faster, hence their sporting popularity.

> **IT'S A FACT...**
> Pheasants spend the night roosting in trees, out of reach of predatory mammals.

An unusual purple-plumaged male Pheasant, captive-bred and released to be shot.

female

female

male

SOCIAL BEHAVIOUR

Most winter groups of Pheasants are of just one sex, with females typically forming larger groups than males. While you usually only see one or two males together, females may gather in groups of four or more, sometimes even in excess of 20. It is interesting to note that there is a clear hierarchy to group structure, something that gains greater focus with the approach of the breeding season. From February, males begin to establish their breeding territories, increasingly displaying to females and directing aggression towards other males. Individual males give a throaty call, coupled with wing-drumming, from a prominent position, to proclaim ownership of a particular territory. A successful male on a good territory will, most likely, attract a number of females, each of which will mate with him, while a male on a poor quality territory may fail to attract a mate altogether. Incubation and the rearing of chicks is the sole responsibility of the hen, who faces a difficult job in our relatively open landscapes where nesting cover may be limited. Nesting on the ground presents obvious problems, not least the potential for predators to discover the young and it is no surprise that the young are born well-advanced, quickly able to walk and to leave the vicinity of the nest with their mother. The young chicks may be tended for 70-80 days before they become independent.

OTHER PHEASANTS

A number of other pheasant species have occurred in Britain over the years, the result of aviary escapes or deliberate introductions. One of these, the Golden Pheasant, has been recorded from Garden BirdWatch gardens and has small feral populations in a number of localities. The male Golden Pheasant is a truly striking bird; however, like other pheasants, the hen is rather drab in appearance, useful camouflage when nesting on the ground.

Male Golden Pheasant.

SUBTLE VARIATION

With his iridescent head, red fleshy 'goggles' and bronze body, the male Pheasant is a stunning bird. Many males show a white collar, a feature derived from the 19th-century introduction of a hardy Chinese race, known as *torquatus*, which has largely replaced the *colchicus* race introduced from the Caucasus. However, with 30 races recognised worldwide, many British pheasants show a very mixed ancestry.

STATUS

Green listed and population reinforced through continual release of hand-reared birds for shooting.

'Ring-necked' male Pheasant.

SEASONAL TRENDS

BTO *Garden BirdWatch* REPORTING RATE (%) THROUGHOUT THE MONTHS OF THE YEAR

BTO BIRDFACTS

BLACK-HEADED GULL
Larus ridibundus

FAMILY
Laridae

LENGTH
36cm

WINGSPAN
105cm

WEIGHT
Male 330g;
female 250g

HABITAT
Lakes, rivers, moors,
grassland and
coasts.

FOOD
Opportunistic, taking
insects, earthworms,
plant material and
scraps.

NEST
A shallow scrape or
raised mound on wet
ground, lined with
vegetation. Usually
in low vegetation but
may be exposed.

EGGS
2-3 eggs, very
variable in colour,
ranging from dull
grey to cinnamon-
brown and covered in
darker brown or olive
blotches.

EGG SIZE
51.9 × 37.2mm

INCUBATION
23-26 days. 1 brood.

BREEDING SEASON
April to August

POPULATION SIZE
128,000 breeding
pairs; young and
winter immigrants
swell the population
to 2.2 million
individuals in winter.

TYPICAL LIFESPAN
6 years

MAX. RECORDED AGE
27 years 1 month

The Black-headed Gull is the most frequently reported gull species to visit garden feeding stations. Despite the name, Black-headed Gulls don't have a black head. During the breeding season it is a dark chocolate brown colour, while in the winter the head is white, with a dark brown smudge on the side.

A CHANGE IN FORTUNES

This species has seen a remarkable change in fortunes over the last 200 years, recovering from near extinction in Britain during the 19th century, with dramatic population increases during the 20th century. Although some inland colonies may consist of just a few breeding pairs, some really big colonies can top 10,000 pairs. These increases were accompanied by a rise in the use of inland areas for breeding, roosting and feeding, to the extent that birds started to visit bird tables in large urban areas like London – the first individuals wintered in central London just 100 years ago. Black-headed Gulls most frequently appear in gardens during the cold winter months of January and February but are increasingly visiting gardens throughout the year, with dispersing birds often present during June and July. Outside of the winter, these garden visitors are usually immature, non-breeding birds from local colonies, recognisable by their brown wing feathers. Since their modern-day peak in the 1970s, Black-headed Gull numbers have fallen back somewhat and this has prompted their addition to the Amber list of birds of conservation concern.

EASTERN EUROPEAN IMMIGRANTS

During the winter, our largely resident breeding population of Black-headed Gulls is joined by large numbers of immigrants from breeding populations scattered around the Baltic, the Low Countries and from as far east as Russia. It has been estimated that more than two-thirds of the birds wintering here may be immigrants, arriving over a protracted period from July to February. The earliest to arrive are usually young birds from breeding colonies in the Low Countries, while the last to arrive may be those forced by freezing weather to leave wintering grounds farther east. Once here, most of the winter visitors appear to be quite site faithful, even to the extent of returning to the same wintering sites in subsequent years. The weather conditions in Britain and Ireland during the winter are mild in comparison with much of northwest Europe, making us an attractive destination for wintering birds.

winter adult

summer adult

IT'S A FACT...
Daphne du Maurier's short story 'The Birds' was inspired by a throng of Black-headed Gulls.

AGE-RELATED PLUMAGE CHANGE

juvenile — 1st winter — 1st summer — adult winter — adult summer

CHEEKY SCAVENGERS

Being scavengers, Black-headed Gulls will take a very wide range of kitchen scraps, from meat and bread to cheese and potatoes. When feeding, they tend to be very aggressive and groups often squabble over the available food. Sometimes, particularly where food is presented in a relatively confined space, the birds don't land but instead dive down to snatch larger food items. Individuals appear bolder in cold weather, no doubt a reflection of the lack of feeding opportunities elsewhere.

FROM DINNER TABLE TO BIRD TABLE

While the Black-headed Gull is no longer seriously regarded as a food item, it was once valued as such. The large breeding colonies were heavily exploited for both their eggs and meat throughout several centuries, with commercial exploitation continuing right up until the 1940s. The sheer scale of the harvest can be gauged by the fact that a large colony might provide some 18,000 eggs per season, many of which would have been transported around the country or else used locally in the making of puddings. Even in the 1930s, Leadenhall Market, situated between Gracechurch Street and Lime Street in London, was still handling a trade of nearly 300,000 eggs per year.

WHAT'S IN A NAME?

The Latin name used for the bird that we know as the Black-headed Gull is *Larus ridibundus*. While *Larus* is the generic name used for a group of related gull species, *ridibundus* literally translates as 'laughing' and refers to the Black-headed Gull's call. Interestingly, a related species, the Laughing Gull of North America, has the Latin name *Larus atricilla*, which can be translated as 'Black-headed Gull', in this instance referring to the black head of the Laughing Gull. Just to confuse things further, our Black-headed Gull has a chocolate brown head, rather than the black one alluded to in its English name!

IT'S A FACT...

Black-headed Gulls visiting winter gardens will quite often show a small smudge behind the eye and the brown wing feathers characteristic of a young bird (*seen here*).

1st winter

adult summer

STATUS

Amber listed.

SEASONAL TRENDS

BTO *Garden BirdWatch* REPORTING RATE (%) THROUGHOUT THE MONTHS OF THE YEAR

BTO BIRDFACTS

FERAL PIGEON
Columba livia

Referred to by some as a 'rat with wings' but treasured by others, the Feral Pigeon is a surprisingly interesting, resourceful and successful bird. The large flocks that are such a feature of many urban centres may despoil our architectural heritage but fascinate children who may not have such close access to other free-living wildlife.

FAMILY	Columbidae
LENGTH	32cm
WINGSPAN	66cm
WEIGHT	300g
HABITAT	Built-up areas.
FOOD	Scraps of many kinds.
NEST	A loosely constructed shallow cup of stems, leaves and other material.
EGGS	2 eggs, glossy white.
EGG SIZE	39.3 × 29.1mm
INCUBATION	16–19 days. 5 or more broods.
BREEDING SEASON	Able to breed all year round
POPULATION SIZE	In excess of 100,000 pairs.
TYPICAL LIFESPAN	3 years
MAX. RECORDED AGE	No accurate data available for Feral Pigeon (Rock Dove max. recorded age, 7 years 1 month)

A WILD ANCESTRY

The familiar Feral Pigeon is descended from the Rock Dove, a species that has been domesticated for many thousands of years. It is now just about impossible to disentangle truly wild Rock Doves from their feral cousins, with birds of feral origin often occupying cliff-nesting sites that once held their wild counterparts. Within Britain and Ireland, truly wild Rock Doves are now confined to remote sea-cliffs in the extreme north and west. Many of the Feral Pigeons breeding in urban areas have plumage that closely resembles that of their truly wild relatives, although there is a great variety of plumage forms, thanks mainly to the continued escape of birds from dovecotes and the thousands of racing pigeons that go astray each year. Intriguingly, melanistic birds appear to have more broods during the course of a year (sometimes five or six) than birds with 'wild type' plumage. This may explain why dark morphs often dominate populations living within our towns and cities.

IT'S A FACT...

Pigeon meat has long been considered good-eating, with nestlings especially prized.

IT'S A FACT...

The hugely popular hobby of pigeon racing provides an important source of recruitment into Feral Pigeon populations.

ON THE MENU

While the spread of the Feral Pigeon largely stems from its value to us as a source of protein, it is no longer harvested in the way it once was. However, urban populations of Feral Pigeons are now exploited by the increasing population of urban-nesting Peregrines. In some small way, the humble pigeon is, inadvertently, doing its bit for conservation!

LITTLE STUDIED

Despite its widespread distribution and association with Man, the Feral Pigeon has been little studied. What work that has been done, specifically that carried out in Bristol by Garden BirdWatcher John Tully, has revealed that urban pigeon populations are pretty much maintained by the food provided directly (or accidentally) by the human population. The 7,440 pigeons found in John's Bristol study area were taking 0.22 tonnes in food each day and, from this, it could be calculated that each pigeon needed 67 humans to support it through the winter.

COLOUR FORMS

Varying degrees of albinism and melanism occur within Feral Pigeon populations and a wide range of plumage patterns can be seen. Many individuals show signs of injury or disease, the latter often resulting from the crowded conditions under which they live.

Ornamental bird displaying.

Pure white 'dove'.

URBAN PROBLEMS

Urban populations of Feral Pigeons have the capacity to increase rapidly, a consequence of their breeding ecology (young can breed at seven months of age), coupled with the favourable conditions in our unhygienic urban centres. Much money is spent to try to control Feral Pigeon numbers and thereby to remove the risk of disease transmission and reduce damage to buildings from the birds' droppings. Restricting access to favoured perches or nest sites may solve the problem at the local level but it simply forces birds to move elsewhere. Culling is largely ineffective because there is a large proportion of non-breeding birds to take over vacated nest sites. The only workable solution seems to be to remove their food supply, by changing peoples' attitudes towards the feeding of pigeons and the disposal of food waste.

A GARDEN NUISANCE?

Feral Pigeons can become a nuisance in gardens where kitchen scraps, bread or cheap bird food (full of wheat or other cereal grains) are provided. Since Feral Pigeons prefer to feed on the ground, the attractiveness of a garden feeding station can be reduced by only providing good quality feed (e.g. black sunflower seed) in hanging feeders. Ground feeding can continue if you can exclude the pigeons from the feeding area by surrounding it with a 2-inch gauge wire mesh. There are various feeder 'guardians' designed specifically for this purpose, which can be purchased from bird food suppliers and garden centres. The Garden BirdWatch reporting rate for this species averages about 10%, peaking in spring when food may be less abundant elsewhere. There has been a slight but steady increase in the Garden BirdWatch reporting since the project began in 1995.

Feral Pigeon plumage is variable but this bird's plumage is similar to that of its ancestor, the Rock Dove.

STATUS

Not a species of conservation concern, hence Green listing.

SEASONAL TRENDS

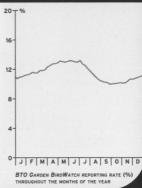

BTO GARDEN BIRDWATCH REPORTING RATE (%) THROUGHOUT THE MONTHS OF THE YEAR

BTO BIRDFACTS

WOODPIGEON
Columba palumbus

FAMILY	Columbidae
LENGTH	41cm
WINGSPAN	78cm
WEIGHT	450g or more
HABITAT	Woodland, farmland and, increasingly, urban areas.
FOOD	Seeds, buds and other plant material.
NEST	A simple and rather pathetic platform of twigs placed on a branch or in the fork of a tree.
EGGS	2 eggs, white in colour and slightly glossy.
EGG SIZE	44.1 × 29.8mm
INCUBATION	17 days. 1-2 broods.
BREEDING SEASON	March to October (though can be longer)
POPULATION SIZE	At least 3 million breeding pairs.
TYPICAL LIFESPAN	3 years
MAX. RECORDED AGE	17 years 8 months

The Woodpigeon is becoming increasingly familiar in many of our urban centres, a result of an increasing population now overspilling from farmland into other habitats. While individuals that have taken to urban living are bold, their country cousins are much more wary. This is hardly surprising given that rural pigeons are still shot in large numbers because of the economic damage that they do to crops.

FEEDING AND BREEDING

For some garden birdwatchers, the Woodpigeon is an unwelcome visitor. It is easy to understand why, since one Woodpigeon can eat as much food as seven sparrows. The Woodpigeon boasts the dubious reputation of being the most economically damaging bird species in Britain and Ireland. Its pest status has increased as it has adapted to new agricultural crops, like oilseed rape, which provide Woodpigeons with a ready supply of green foodstuffs during the difficult winter months. Set against such problems with its image, the Woodpigeon remains one of the most biologically interesting of our garden birds. For example, the Woodpigeon is one of a very small number of birds that produce 'crop milk', a substance that is very similar to mammalian milk in its composition. The milk is produced, as the name suggests, in the crop, which is a sac-like structure normally used for storing food. During the last few days of the incubation period some of the cells lining the crop begin to produce milk, which forms the only source of nourishment during the first few days of young birds' lives. Other nestlings of seed-eating species have to be fed on animal material during their period of maximum growth.

adult

IDENTIFICATION

An adult Woodpigeon is a large, plump bird with a white neck patch. There is a smudge of petrol-green above this. A young bird also has the patch of green but lacks the white of the adult. The 'coo-coo-coo-co-co' song, often heard during April and May, is a familiar part of the dawn chorus to most garden birdwatchers, and can be annoying if the bird happens to be calling from your chimney. The easiest way to learn how to identify the characteristic call (you cannot really call it a song) is with the rhyme 'take-two-cows-taffy'.

AUTUMN BOUNTY

Woodpigeons take advantage of the autumn crop of acorns and beechmast, congregating in large flocks where these tree seeds happen to be plentiful. In years when these seeds are scarce, some Woodpigeon populations may undertake migrational movements. Such movements are more common in continental populations than those residing here.

adult

BREEDING BEHAVIOUR

Woodpigeons can be gregarious in nature outside the breeding season but during spring the males set up small breeding territories. Ownership of these is advertised through the display flight – the male soaring upwards before stalling and executing one or two wing claps. The nest is a rather flimsy lattice of sticks, placed on a branch or in the fork of a tree or shrub. Sometimes the nest is so poorly constructed that you can actually count the eggs from below! Mind you, since Woodpigeons invariably lay two eggs (one for each of the two incubation patches on either side of the female's breast), there is no real need to count them in the first place. The breeding season itself is very long, with eggs having been found in every month of the year, although the main period of activity is over by the end of September.

A Woodpigeon on its flimsy nest in early spring.

Birds in flight, showing the striking white wing and neck patches.

Adult Woodpigeons often feed on lawns.

IT'S A FACT...

A Woodpigeon is a surprisingly heavy bird; weighing 450g it is equivalent to the weight of four and a half Blackbirds or 41 Blue Tits! No wonder that they can consume so much food.

Urban Woodpigeons often use walls and buildings as lookouts and roosting spots.

STATUS

Increasing, hence Green listed.

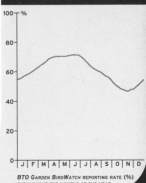

SEASONAL TRENDS

BTO GARDEN BIRDWATCH REPORTING RATE (%)
THROUGHOUT THE MONTHS OF THE YEAR

THE DAILY GRIND

Woodpigeons have a diet that includes a mixture of seeds and other plant material. Seeds are swallowed whole, passing into the gizzard – a muscular grinding organ – where they are broken down through abrasion. This process is helped by the presence of small pieces of grit, deliberately swallowed by the bird for this purpose. Woodpigeons are also able to store a certain amount of food in their crop, allowing them to feed quickly and then digest the food in the relatively security of their roost.

URBAN NESTING

Within urban areas, Woodpigeons usually nest in trees or tall shrubs but they have also been recorded nesting on ledges on buildings. The Garden BirdWatch reporting rate shows annually increasing use of gardens over time. Woodpigeons were first noticed colonising urban areas some 180 years ago and they can often be seen feeding alongside Feral Pigeons. Many of these urban Woodpigeons feed on cheap seed mixes and on kitchen scraps, notably bread, and this may allow them to begin breeding earlier in the year than their rural cousins.

BTO
BIRDFACTS

COLLARED DOVE
Streptopelia decaocto

This is a bird which has taken Britain by storm, exploiting a vacant niche to become one of our most familiar garden birds. Breeding throughout the year and making the most of garden feeding has certainly helped the Collared Dove population to increase rapidly.

FAMILY
Columbidae

LENGTH
32cm

WINGSPAN
51cm

WEIGHT
200g

HABITAT
Gardens, scrub and open woodland.

FOOD
Mostly cereal grain, also weed seeds, occasional shoots and invertebrates.

NEST
A flimsy platform of twigs, placed on a ledge or in the fork of a tree.

EGGS
2 eggs, glossy white in appearance.

EGG SIZE
31 × 24mm

INCUBATION
16-17 days.
3-6 broods.

BREEDING SEASON
February to October

POPULATION SIZE
284,000 pairs.

TYPICAL LIFESPAN
3 years

MAX. RECORDED AGE
16 years 10 months

COURTSHIP AND DISPLAY

Males defend small territories around the nest by keeping watch for other males from a suitable vantage point. In addition, a male will advertise ownership of a territory by giving a display flight. Like the Woodpigeon this consists of a steep climb, followed by a glide, though the glide is often spiral in nature in the Collared Dove, with the tail well spread. An excitement call may sometimes be uttered and this has been described as a rather jarring 'rrräh'. The more familiar, and rather monotonous, 'ku-koo-ku' call also has a territorial function and is often unpopular with people attempting to sleep through the dawn chorus!

YEAR-ROUND BREEDING

Another factor in the success of the Collared Dove has been its breeding ecology, in particular its long breeding season and the tendency to start a new breeding attempt before the previous one has been completed. There are instances reported where incubating females use their 'off duty' breaks from incubation to attend to the recently fledged young of the previous brood. The main breeding season within Britain extends from mid-February through to early October, but they have been reported nesting in Christmas trees before Twelfth Night. The nest, a pathetic platform of twigs, is usually placed in a tree or shrub but may sometimes be built on a ledge or similar structure. Pairs nesting in suburban habitats are, on average, less successful in their breeding attempts than those breeding in rural areas. However, the sheer number of breeding attempts made during the course of the breeding season results in the production of enough fledged young to ensure that the population keeps increasing.

IT'S A FACT...

Over recent years the Collared Dove has been the most commonly reported of our pigeons and doves, making widespread use of gardens. However, the recent increase in Woodpigeon numbers may just nudge it off the top spot.

HOW TIMES HAVE CHANGED

It is amazing to think that the now ubiquitous Collared Dove was first recorded breeding in Britain during the 1950s. This colonisation of Britain and Ireland was part of a larger range expansion that took place across much of Europe. Before 1930, Collared Doves were pretty much restricted, within Europe, to Turkey and the Balkans. For some reason, possibly linked to a genetic change in the dispersal behaviour of young birds, the population began to push northwest across Europe. In just 20 years, the species spread rapidly, covering a distance of more than 1,600km. The first breeding record in Britain came from Norfolk in 1955 but very quickly other areas were colonised. Most of the initial spread was linked to rural villages and town suburbs and it is interesting to note that this species shows a stronger association with Man here than it does elsewhere in its range. Of particular importance appears to be access to grain, either spilt on farmland or provided at suburban bird tables. The importance of cereals in the diet also explains the large flocks that gather outside the breeding season. Having said this, Collared Doves will also make use of other plant material, like seeds and berries, and will feed as happily on garden plants as they do on arable weeds.

In Germany, the Collared Dove has become known as the 'television dove' because it commonly calls from rooftoop aerials.

ON THE MOVE

Although we tend to think of Collared Doves as being pretty sedentary, young birds may move some distance away from where they were born, with some travelling more than 600km. The movements made by young birds show a tendency to be northwest in their orientation. This pattern is repeated elsewhere in Europe and reflects the underlying direction of the range expansion, from Turkey and the Balkans northwest across Europe.

LIFE AND DEATH

Something like two-thirds of young Collared Doves will die before the end of their first year of life. Inexperience, competition for food, predators and disease will all take their toll. Sparrowhawks and cats often take Collared Doves and this may be one reason why Collared Doves increasingly favour dense conifers and holly bushes for roosting.

URBAN COMPANION

This medium-sized dove has a rather sleek appearance, except when fluffed up against the cold. Generally buff-grey in colour, adult males may show a soft pink flush to their plumage. The dark eye, white eye ring and unique black collar are useful identification features.

An adult Collared Dove confidently strutting on a garden shed roof.

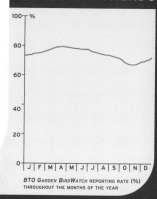

BTO BIRDFACTS

TAWNY OWL
Strix aluco

FAMILY	Strigidae
LENGTH	38cm
WINGSPAN	99cm
WEIGHT	Males 420g; females 520g
HABITAT	Woodland and wooded suburbs.
FOOD	Small mammals, small birds, worms and amphibians.
NEST	Large cavity or nestbox, usually some height above the ground in a mature tree.
EGGS	2-3 rounded white eggs, slightly glossy.
EGG SIZE	46.7 × 39.1mm
INCUBATION	30 days. 1 brood.
BREEDING SEASON	February to June
POPULATION SIZE	19,000 breeding pairs.
TYPICAL LIFESPAN	5 years
MAX. RECORDED AGE	21 years 5 months

IT'S A FACT...

Owls used to be viewed as omens of doom, their hooting calls said to portend bereavement in the house.

This is the owl most likely to be reported using large gardens, churchyards and parks, reflecting its ability to exploit a wide range of prey species. Tawny Owls are absent from Ireland, the Isle of Man and most of the other islands around our coast.

ROOSTING AND FEEDING

Tawny Owls generally spend the daylight hours roosting in trees, often against the trunk or amongst ivy. This makes them difficult to spot, so they only tend to be recorded when their characteristic 'tuweet' and 'tu-woo' calls are made. The males and females utter a range of different calls, though both make the 'ke-wick' contact call that can often be heard through the dark winter months. Calling rates of territorial birds are influenced by weather conditions, the owls calling less on wet or windy nights when their calls may not carry as far. Tawny Owls are primarily woodland birds, feeding on small mammals and occasionally taking small birds. In urban areas there are not the numbers of mice and voles to support them, so brown rats, grey squirrels, Starlings and House Sparrows are eaten more often. These are taken at night, the owl using its very sensitive hearing and vision to drop on a prey item from a favoured perch or to dive into bushes occupied by roosting birds. Tawny Owls will also hunt on foot, coming down to feed on earthworms in much the same way as Blackbirds do.

adult

CAVITY NESTER

Tawny Owls need to find a suitable cavity in which to nest. This is one reason why they are an uncommon visitors to most gardens; the reporting rate from Garden BirdWatch gardens is very low, averaging about 3%, with a peak from late autumn through into the winter, when the owls are very vocal in defence of their territories. Cavities used for nesting are often surprisingly open and this may explain why Tawny Owls will also take over the large stick nests of crows and Sparrowhawks. Tawny Owls will also utilise specially designed nestboxes, erected high on the trunk of a tree or slung from a strong branch. Young Tawny Owls leave the nest well before they are able to fly. They clamber about in the branches and occasionally fall to the ground, where they are sometimes found by garden birdwatchers. Unless the youngster is in immediate danger of being taken by a cat, it is best to leave it where found. Youngsters are quite capable of climbing back up tree trunks by using their incredibly strong claws.

juvenile

DESIGNED FOR WOODLAND LIVING

If you are lucky enough to see a Tawny Owl, then you will notice that it has short, broad wings and tail, features that help it to manoeuvre through its preferred woodland habitat. Owls that hunt on the wing in open country, like the Barn Owl, have longer wings. The Tawny Owl's plumage is a streaky brown colour that provides camouflage, helping it to remain unnoticed at its daytime roost. Tawny Owls are strongly territorial, a behaviour that enables them to develop a good knowledge of where suitable prey can be found and the best perches from which to hunt them. The large black eyes are superbly adapted to hunting at night; even in the dark of night the owl is able to swoop down with pinpoint precision onto unsuspecting prey.

Roosting Tawny Owl being mobbed by garden birds.

GOOD PARENTS

Adult females are very good parents and can be aggressive at the nest. A well-known wildlife photographer, the late Eric Hosking, lost an eye to a Tawny Owl early in his career. Evidence suggests that Tawny Owls nesting in well-populated areas may be more aggressive than those breeding in more remote locations. All in all, nesting Tawny Owls are best left well alone.

IT'S A FACT...

The amazing facial disk enables the owl to direct sound waves onto its ears, which lie hidden beneath the feathers on either side of the disk. The two ears are slightly out of alignment with each other, giving the owl exceptional directional hearing.

STATUS

Slight downward population trend but not sufficient at this stage to upgrade conservation status from Green listing.

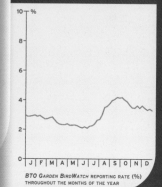

PELLETS

The bones from the Tawny Owl's small mammal meals are indigestible. Rather than risk damaging itself by passing these bones through its gut the owl coughs them back up, wrapped in fur and other indigestible material. One or two pellets may be produced each day and these can be collected and dissected by those wishing to study the owl's diet. Soak a pellet in water for a couple of hours and then pull it apart to find the bits of bone and small mammal skulls. With the right field guide it is possible to identify the different sorts of small mammal that the owl has been feeding on, including mice, voles and shrews.

Tawny Owl with vole prey   **pellet**   **vole bones extracted from pellet**

SEASONAL TRENDS

BTO GARDEN BIRDWATCH REPORTING RATE (%) THROUGHOUT THE MONTHS OF THE YEAR

BIRDFACTS

FAMILY	Strigidae
LENGTH	22cm
WINGSPAN	56cm
WEIGHT	180g
HABITAT	Farmland and open country.
FOOD	Small mammals and small birds, large invertebrates.
NEST	Cavity in tree, wall or suitable nestbox, may use an burrow.
EGGS	3-4 eggs, white in colour.
EGG SIZE	35.6 × 29.6mm
INCUBATION	29-31 days. 1, sometimes 2 broods.
BREEDING SEASON	March to July
POPULATION SIZE	7,000 breeding pairs.
TYPICAL LIFESPAN	3 years
MAX. RECORDED AGE	10 years 2 months

IT'S A FACT...

In addition to the small size, the other features to look out for are the bright yellow eyes, oblique white eyebrows and strongly undulating flight.

LITTLE OWL
Athene noctua

This, as the name implies, is the smallest of our owls; not much bigger than a Song Thrush in size, with a compact body and a flat-topped head. Little Owls are only occasional visitors to gardens, sometimes using those that are large and rural in nature.

adult

INTRODUCTIONS

Despite the fact that the Little Owl is now accepted as being a legitimate part of our avifauna it is not regarded as being native. Instead the widespread breeding population is thought to be descended from birds that were imported and then released here during the last quarter of the 19th century. The native range of the species actually extends across mainland Europe, from Denmark southwest into North Africa and east across the Middle East, Asia and across to the Pacific coast. The first documented attempts to introduce the Little Owl into Britain, using birds imported from Italy, were made in the 1840s but these attempts failed, as did four others made to other parts of England. The first successful introductions were made to Edenbridge in Kent and Oundle in Northamptonshire between 1870 and 1880. These resulted in the first recorded breeding attempts just a few years later and from there the population expanded rapidly, such that most of the current British range was occupied by the 1920s. Since then numbers have peaked and then declined, a reflection of a changing environment, the loss of favoured nesting sites and a decline in the abundance of favoured prey species. This decline matches a pattern seen elsewhere across Europe and has given cause for concern.

adult

adult

PELLETS

Like other owls, and indeed many other birds, the Little Owl regurgitates indigestible prey remains in the form of pellets. These are fairly small in size, typically 25 × 14mm and usually rounded at each end. Often grey-brown in colour, the pellets contain small mammal bones and, characteristically, a large number of beetle wing cases. Those individuals that have been feeding on earthworms produce pellets with the consistency and colour of sand; the 'sand' being the soil particles left after the earthworm and its stomach have been digested!

ACTIVITY

Although typically active at night and around dusk and dawn, Little Owls may be seen during the daytime, perched in a tree or, occasionally, seen in flight. Perches are often used as somewhere from which to scan for suitable prey, the owl dropping down onto an unsuspecting small mammal, small bird or large invertebrate. Despite its size the Little Owl is quite capable of tackling a prey item up to the size of a half-grown Rabbit, though it is equally at ease pursuing a large beetle across the ground, or taking earthworms brought to the surface on a damp night. Some individuals may specialise in taking particular prey. One pair, nesting on the island of Skokholm (off the coast of Pembrokeshire), fed almost exclusively on Storm Petrels (diminutive seabirds that nest on the island).

BREEDING

Although the breeding season can begin as early as January, it is at its peak during March and April when the males proclaim ownership of their territories through a 'hoo-hoo, hoo' call, answered by the female through a series of short shrieks and yelps. The Little Owl has quite an impressive vocal repertoire and, in addition to the territory calls, a plaintive 'kiew kiew' call may often be heard. The breeding pair will spend increasing amounts of time around the chosen nest site, the male demonstrating its suitability to his mate and also presenting her with choice items of prey. These behaviours serve to cement the pair bond and strengthen the association with the nest site. The three or four eggs are almost invariably laid during April and May, the owls favouring a small cavity in an old deciduous tree. Some pairs will nest in cavities within the walls of an old farm building while others may utilise the abandoned burrow made by a Rabbit. If the spring weather is particularly poor then the pair may not attempt breeding that year.

BOUNDARIES TO DISTRIBUTION

Large expanses of water are an obstacle to dispersing Little Owls. This reluctance to cross such waterbodies goes a long way to explaining why just a handful of individuals have ever reached Ireland.

IT'S A FACT...

Anxious Little Owls may be seen bobbing up and down in a rather comical fashion.

adult

IT'S A FACT...

Young Little Owls may sometimes produce a hissing sound, particularly when they are excited or anticipating a feeding visit from one of their parents.

STATUS

Possible decline in Britain but not listed as being of conservation concern.

SEASONAL TRENDS

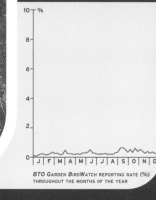

BTO GARDEN BIRDWATCH REPORTING RATE (%) THROUGHOUT THE MONTHS OF THE YEAR

Adult with young at nest.

Nest sited in an abandoned shed.

BTO BIRDFACTS

SWIFT
Apus apus

FAMILY	Apopidae
LENGTH	16cm
WINGSPAN	45cm
WEIGHT	44g
HABITAT	Open country and towns.
FOOD	Small insects and spiders caught in flight.
NEST	Shallow cup of grass, leaves and feathers, cemented together with saliva and placed on a flat surface under eaves of roof or in a hole in a wall.
EGGS	2-3 eggs, white in colour.
EGG SIZE	25.0 × 16.3mm
INCUBATION	19-25 days. 1 brood.
BREEDING SEASON	May to August
POPULATION SIZE	80,000 pairs.
TYPICAL LIFESPAN	9 years
MAX. RECORDED AGE	17 years 11 months

There is no other bird that signals the arrival of summer in our towns and cities as evocatively as the Swift, whose spring arrival is sudden and breathtaking – the sickle-winged, screaming silhouettes that dash about the sky.

WHERE TO NEST

Within Britain and Ireland, nests are placed under eaves in a cavity in a building or in a specially designed nestbox. In parts of eastern Europe, Swifts prefer tree cavities or crevices in cliff faces. Old buildings are favoured over more modern constructions because modern houses lack the access under the eaves that Swifts require. Swifts seem to favour tall buildings because the extra height allows them to drop from the nest site and reach sufficient speed to get airborne. Grounded Swifts find it particularly difficult to get into the air and consequently healthy birds never willingly land, apart from when visiting the nest. Non-breeding birds roost on the wing during the breeding season, gathering together in an evening ascent that takes them to higher altitudes, where they appear to sleep whilst using a gliding flight.

NEST STUDIES

Work on nesting Swifts, specifically in Oxford and Portsmouth, has provided some unique insights into the breeding ecology of this species.

Young Swifts in the nest.

BREEDING AND DEPARTURES

The first pairs to breed will typically lay three eggs, those starting later just two. The weather conditions, and hence the availability of aerial insects, determines not only when breeding starts but also how long it takes. During wet or windy conditions, Swifts may find it difficult to find food for their developing chicks. The chicks can cope with this to some extent by going into a torpor, which may last for several days if the weather is bad. The consequence of this is that chicks can take between 35 and 56 days to fledge from the nest. Once they leave, they are fully independent and depart for their wintering grounds in Africa within a few days. One ringed chick that left its nest in Britain was killed in Madrid four days later. It is amazing to think that once a chick has left its nest it will not stop flying again until 23 or more months later when it first prospects for a breeding site of its own. This means that the fledgling's first flight is likely to cover some 570,000km! The adults remain for a few days longer and then depart south. Although the 2 birds that make up a breeding pair may have no contact with each other outside of the breeding season, they invariably return to the same breeding site, maintaining the pair bond from one year to the next.

Adult Swift incubating its eggs.

Depending on whether they are soaring or actively feeding, the profile of a flying Swift is rather variable.

Swifts even drink on the wing.

IT'S A FACT...

Young Swifts are fed on compacted balls of food, mostly flying insects, caught by the adults as they forage, often many kilometres from the nest.

SUMMER ARRIVALS

Swifts are one of the last of our summer migrants to arrive and first to depart, typically arriving in late April and remaining here for just 16 weeks. Not all of the Swifts that arrive for the summer are here to breed; many are young birds that won't nest for their first few summers. A delay in reaching breeding age is unusual in a small bird because of the high levels of mortality that small birds usually suffer. It should be no surprise, therefore, that only one in six adults die in a given year, a much lower figure than that for similarly sized birds. It is the young, non-breeding, birds that make up the screaming parties of Swifts that charge about our skies during June and July, bombing about at rooftop level like tiny jet fighters, focusing their attention on established colonies. Adults occupying the nest holes scream back as the youngsters fly up and knock repeatedly against nest entrances. Only if all the apparently suitable holes are occupied will the party of young Swifts try its luck elsewhere. Identifying potential nests sites for a future breeding season may seem like long-term planning but it is clearly a strategy that works. This approach does mean, however, that the establishment of new colonies is a slow process. Even when a new colony is established, the arriving birds tend not to nest in their first year of occupancy.

IT'S A FACT...

Even though Swifts have very short legs, these are covered in feathers to reduce heat loss when the birds fly at high altitude.

juvenile

adult

CONCERN FOR SWIFTS

The loss of suitable nest sites, either through the redevelopment of old properties or the renewal of their roofs is a serious problem for our Swifts. The degree of nest site loss has prompted the establishment of a number of voluntary organisations who provide advice on how to retain Swifts when a building is redeveloped. Quite often this involves the use of Swift nestboxes or special Swift bricks.

STATUS

Green listed.

SEASONAL TRENDS

BTO *GARDEN BIRDWATCH* REPORTING RATE (%)
THROUGHOUT THE MONTHS OF THE YEAR

BTO BIRDFACTS

FAMILY
Alcedinidae

LENGTH
16cm

WINGSPAN
25cm

WEIGHT
40g

HABITAT
Slow-moving rivers, lakes and estuaries.

FOOD
Mostly small fish but will also take aquatic invertebrates.

NEST
A tunnel, some 90cm long, excavated in a soft riverside bank.

EGGS
5-7 eggs, almost spherical in shape and white in colour.

EGG SIZE
22.6 × 18.8mm

INCUBATION
20-21 days. 1-3 broods.

BREEDING SEASON
March to September

POPULATION SIZE
6,000 breeding pairs.

TYPICAL LIFESPAN
Not known

MAX. RECORDED AGE
4 years 6 months

KINGFISHER
Alcedo atthis

Although the Kingfisher cannot be described as a regular garden bird, some lucky Garden BirdWatchers report Kingfishers in their gardens every week. Others receive a visit during particularly cold weather, when waterbodies are frozen and the birds seek alternative food elsewhere.

OFTEN OVERLOOKED

Kingfishers spend large amounts of time perched motionless at the water's edge, watching intently for suitably sized fish to pass beneath. This need to spot and reach small fish largely restricts the Kingfisher to our slower-flowing rivers. Birds perched in this manner are often overlooked, despite their bright plumage. However, if disturbed, the lightning blue flash of a Kingfisher moving away is likely to catch the eye.

CHANGING TIMES

Improving water quality has benefited the Kingfisher, by allowing populations of small fish (typically minnows, sticklebacks and small trout) to return to once-polluted streams and rivers. This has enabled Kingfishers to colonise some urban areas, giving the chance for more people to witness the bright blue blur of a Kingfisher disappearing up-river from a favoured perch.

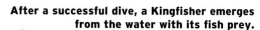

After a successful dive, a Kingfisher emerges from the water with its fish prey.

LITTLE FISHERMAN

Dispersing birds travel cross-country and can turn up at gardens virtually anywhere. Kingfishers may be more regular visitors to garden ponds if these happen to be within a few hundred metres of a larger body of water. The presence of a good stock of small fish and a suitable perch from which to hunt are essential. A perch that is between 1 and 2m above the water is favoured, since this provides an ideal vantage point from which to spot prey. Once the Kingfisher has selected a likely target, it will bob its head up and down to better gauge the position of the fish. The bird will then dive headlong into the water, holding its wings open underwater and protecting its eyes with a special third nictitating eyelid. These dives can take the bird to depths of 25cm, quite deep for a small bird. The Kingfisher then struggles free from the water and will often return to the perch, to perform the delicate juggling act of getting the fish into a head-down position from which it can be swallowed.

Fish prey is usually held tail-first so that it can be stunned on a branch before being flipped and swallowed head-first.

Sited near a garden pond, a tree guard makes a perfectly acceptable perch for a visiting Kingfisher.

NESTING BEHAVIOUR

Kingfishers nest in burrows, typically excavated in soft riverbanks and up to 140cm in length, the tunnel rising slightly along its length to the circular nesting chamber. The construction of such a long tunnel is a difficult task for a bird and takes many days. Several weeks after the eggs hatch, the young emerge and begin a noisy exploration of their surroundings. It is often at this stage that Kingfishers are at their most obvious. Kingfishers invariably raise their second brood in a new nest tunnel, understandable when one considers the stinking mess left in the first nest following a diet largely composed of fish!

ELECTRIC BLUE

The stunning colours of the Kingfisher, electric blue above and rusty orange below, make this one of our most instantly recognisable birds. Despite our familiarity with the species, many observers new to the bird are surprised to discover just how small it is.

IT'S A FACT...

In times past, Kingfishers were sometimes killed and their bodies hung from a thread outside a house in the misguided belief they could accurately predict wind direction.

A male Kingfisher courting his mate with an offering of a fish.

HARD TIMES

During severe winter weather, when water bodies freeze over and Kingfishers are unable to reach their favoured food, some individuals will take kitchen scraps from bird tables in gardens close to rivers, lakes and streams. Birds have been recorded taking offal, suet and even bread, highlighting the tremendous difficulties they face in very cold winters. Results from the BTO's Waterways Bird Survey have highlighted the effect that severe winters, such as 1981/82, can have on the Kingfisher population, sometimes resulting in the local extinction of the species from affected areas.

IT'S A FACT...

The cold days of winter often see Kingfishers moving to the more favourable conditions offered by our coastal estuaries and creeks.

STATUS

Amber listed because of decline in Europe; breeding population is increasing within Britain.

The electric blue feathers on the back of a Kingfisher are simply stunning.

SEASONAL TRENDS

BTO *Garden BirdWatch* REPORTING RATE (%) THROUGHOUT THE MONTHS OF THE YEAR

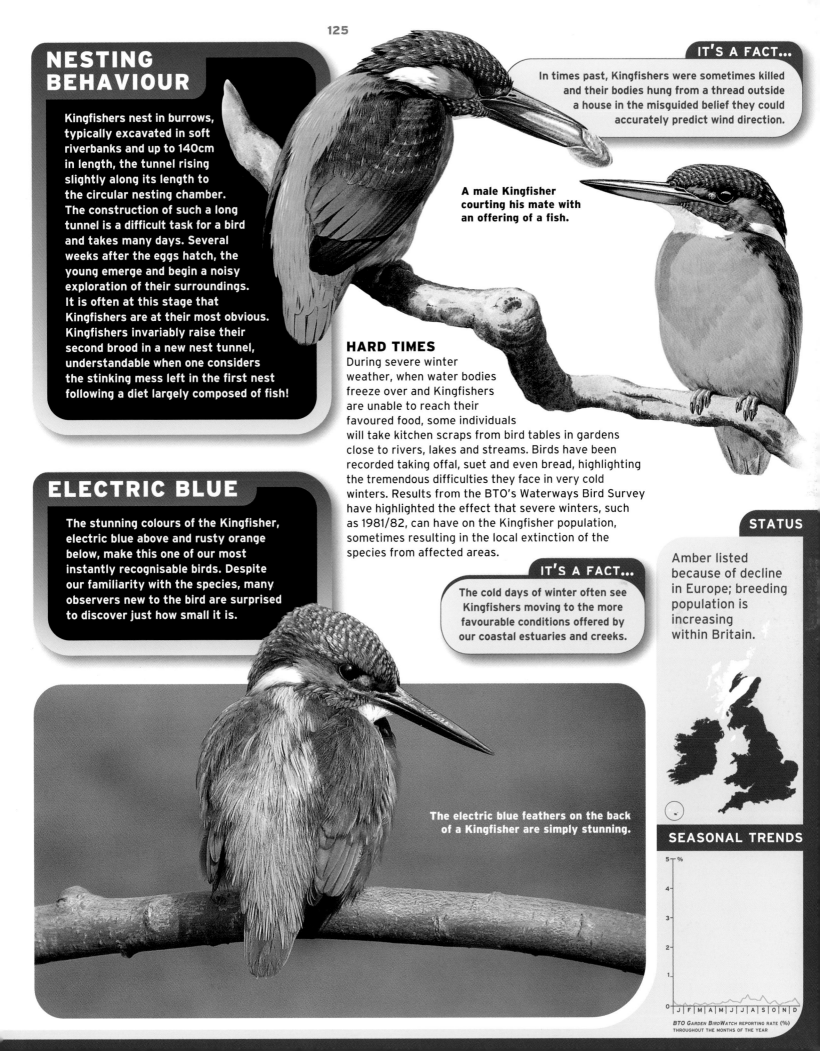

RING-NECKED PARAKEET
Psittacula krameri

The Ring-necked Parakeet is the most colourful species to visit garden bird feeding stations on a regular basis. Unlike many other exotic birds that are occasional escapees from aviaries and private collections, the Ring-necked Parakeet has managed to establish a sizeable breeding population, centred on the Home Counties and the Thames Valley.

TASTE OF THE EXOTIC

Having bright green parrots visit your garden feeding station may come as a bit of a shock, particularly if they then use their sharp beak to rip open your feeders. The Ring-necked Parakeet can no longer simply be regarded as an occasional aviary escape; instead it has become an established part of our avifauna. Breeding in the wild in Britain was first reported in 1969 and the population is now thought to number many thousands of individuals, mainly breeding in larger gardens and parks throughout the suburban parts of Surrey and Kent.

male

female

FAMILY
Psittacidae

LENGTH
40cm

WINGSPAN
45cm

WEIGHT
Males 130g; females 110g

HABITAT
Open woodland, suburbs and parks.

FOOD
Very catholic in its tastes, but mainly feeds on seeds, fruits and flowers.

NEST
Cavity in tree, often old woodpecker hole, some height above the ground.

EGGS
3-4 white eggs.

EGG SIZE
30.5 × 23.7mm

INCUBATION
22-24 days. Thought to have just the 1 brood in Britain.

BREEDING SEASON
January to June

POPULATION SIZE
Breeding population unknown, but wintering population possibly 7,000 plus.

TYPICAL LIFESPAN
Not known

MAX. RECORDED AGE
7 years 7 months

OTHER IMPACTS

Ring-necked Parakeets nest in cavities in trees, taking over holes that have been made by other species and then enlarging them to meet their own requirements. The favoured site consists of a short straight entrance passage that then turns sharply downwards into the larger nesting chamber. This design probably helps to reduce the unwelcome attentions of the nest predators, like snakes, that the parakeet would encounter in its native range. With natural cavities in short supply in Britain, we need to find out how the spread of Ring-necked Parakeets is affecting other hole-nesting species. The fact that these parakeets are cavity nesters puts them in direct competition with species like Little Owl and Jackdaw. Not only are Ring-necked Parakeets aggressive but they also take up residence very early in the year, meaning that they are more likely to occupy potential nest sites before the other competing species get a look in.

WINTER ROOSTS

It is during the early winter months that the size of the Ring-necked Parakeet population can best be assessed. At this time of the year, the birds gather together in large roosts at favoured sites. One such roost, in Surrey, reportedly held a peak of over 7,000 birds during one night. The birds arrive at the roosts before dark, when they are particularly vocal. The parakeets forage in smaller parties during the day and early in the winter favour apples still on the trees. Annoyingly, they never seem to consume whole apples, preferring to take a few pecks before moving on to another. Later in the year, food provided at bird tables and in hanging feeders is important, helping these birds get through the winter. Certainly, severe weather, such as that of the 1981/82 winter, does not seem to have checked their population – our birds derive from the upland Indian race and are adapted to cold nights. It appears that the Ring-necked Parakeet is here to stay, which raises the question of whether it will become a significant pest of major economic importance in orchards. Winter roosts break up soon after Christmas, as individuals begin to pair up, with egg-laying initiated from January onwards.

IT'S A FACT...

Other parrot species have succeeded in breeding in the wild within Britain and these populations also need to be monitored on a regular basis to see if they become established.

PRETTY POLLY

Native to Africa and Asia, Ring-necked Parakeets are attractive birds, yellowish green in colour with a long tail, rosehip red beak and a pink neck collar in the male. Adult birds have the longest tails. In flight, the dark underwing feathers can be seen. The calls made are very loud and squawking in nature. You might imagine that their coloration would make these parakeets easy to spot. However, this is not the case, since the bright green plumage blends in well with vegetation.

PET TRADE

Perhaps unsurprisingly, the parrot family has the largest number of threatened species of any bird family in the world. The main threats are the trade in wild birds and the destruction of their forest habitats.

In flight, Ring-necked Parakeets have powerful wingbeats and a direct path. The dark underwings are noticeable and the long tail makes identification straightforward.

With its neck stretched, the pink elements of this male's neck 'ring' are obvious; the species is sometimes called a Rose-ringed Parakeet.

IT'S A FACT...

There are also small populations of Ring-necked Parakeets breeding in Germany, Belgium and several other countries.

Tail feathers become worn and damaged, and are moulted too. Consequently, some birds appear to have shorter tails than others.

STATUS

Increasing and non-native so not of conservation concern.

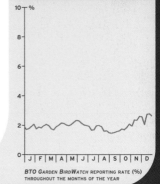

SEASONAL TRENDS

BTO GARDEN BIRDWATCH REPORTING RATE (%) THROUGHOUT THE MONTHS OF THE YEAR

BIRDFACTS

GREEN WOODPECKER
Picus viridis

FAMILY	Picidae
LENGTH	32cm
WINGSPAN	41cm
WEIGHT	190g
HABITAT	Deciduous and mixed woodland, mixed with more open short grassland.
FOOD	Dominated by ants of various species, but will also take other invertebrates, pine seeds and fruits (especially windfall apples).
NEST	Excavated nest hole, situated between 1 and 5m from the ground.
EGGS	4-6 eggs, smooth and glossy white.
EGG SIZE	31.8 × 23.0mm
INCUBATION	19-20 days. 1 brood.
BREEDING SEASON	April to July
POPULATION SIZE	24,000 breeding pairs.
TYPICAL LIFESPAN	Not known
MAX. RECORDED AGE	15 years

The Green Woodpecker, unmistakable with its bright green plumage, yellow rump and red crown, is an occasional visitor to gardens, chiefly to lawns where it feeds on ants but rarely to bird tables.

A DIET OF ANTS

This is the largest of our woodpeckers and spends a great deal of time away from trees, foraging on the ground, where it feeds upon ants, taking adults, larvae and eggs from their nests. To help with this choice of diet, the Green Woodpecker has an extremely long tongue that can be extended to a length of just over 10cm. The tip of the tongue is wide and flat, extremely mobile and covered with sticky saliva produced by the enlarged salivary glands. Green Woodpeckers are often found foraging in areas of old pasture and dry heath, where ant colonies thrive, as well as in woodland glades and on garden lawns. During the summer months, birds seem to take smaller ant species but later in the year they tend to take larger more conspicuous species, often exploiting other habitats like golf courses and parks. In the winter ants can be difficult to find, especially when there is deep snow cover or hard frosts. As a consequence, during particularly bad winters the Green Woodpecker population may decline dramatically.

A WEAK BILL

Young Green Woodpeckers are dependent on their parents for food while in the nest – their diet is mainly insects.

The bill of the Green Woodpecker is weaker than that of other woodpecker species. It is usually only used to chisel on soft wood, which explains why Green Woodpeckers do not often drum on tree trunks. They make up for this by being particularly vocal; the very loud, laughing call gives the bird its local name of 'yaffle' and is a familiar sound to most users of the countryside.

IT'S A FACT...

Green Woodpeckers do not usually take food provided at bird tables or feeders but they have been recorded taking fat, fruit and mealworms provided on the ground.

juvenile

male

female

BREEDING BEHAVIOUR

Green Woodpeckers are monogamous and both members of the pair seek out suitable nesting opportunities, proudly showing these to their mate. In the event, usually more than one nest-cavity is excavated, with the male doing the bulk of the work. Those cavities started by the female often remain unfinished as she increasingly turns her attention to helping the male with what will become the chosen site. Once he has completed the bulk of the work on the chosen site, the female may add a few finishing touches of her own. Since the completed nest site may be a precious commodity, one member of the pair will remain in the cavity, guarding it before nesting proper begins. As with a number of other bird species, the male further cements his bond with the female by presenting her with small items of food prior to mating. As he offers the food to his mate the male will bob his head and sway his body, the female responding to this by begging for food in a similar manner to that seen when a chick begs food from its parent. The female will then adopt a submissive posture which triggers the male to mate with her. Unlike other woodpeckers, the Green Woodpecker may sometimes mate on the ground, reflecting its greater ease on the ground than other woodpecker species.

adult female

DISTRIBUTION

Although the Green Woodpecker has been extending its range in Britain (breeding for the first time in Scotland in 1951), it is absent from Ireland (there are just three very old records) and the Isle of Man, and is most abundant in the southeast of Britain. During a 20-year period of range expansion, the species spread some 200km further north, despite being a rather sedentary species. Green Woodpeckers also appear reluctant to cross large water bodies, the Isle of Wight not being colonised until 1910.

male

IT'S A FACT...

Although Green Woodpeckers pair for life, they are surprisingly antisocial and spend most of the year living a solitary existence. At best, the male and female maintain a loose bond outside of the breeding season.

MALES & FEMALES, YOUNG AND OLD

If you spot a Green Woodpecker in your garden over the summer have a go to see if you can work out its age and sex. In late summer, young Green Woodpeckers have a small amount of grey showing in the red of their crown, while the adults have pure red crowns. The males have red moustachial stripes under the black patch that surrounds the eye. In young males the amount of red in the moustachial stripe is reduced and more black feathering shows in and around the stripe, while in females it is completely lacking in red.

adult female

juvenile

STATUS

Increased steadily in Britain over the last 50 years but amber listed because of status elsewhere in Europe.

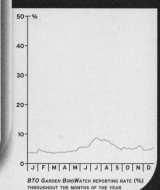

SEASONAL TRENDS

50 %
40
30
20
10
0

J F M A M J J A S O N D

BTO *Garden BirdWatch* reporting rate (%) throughout the months of the year

COMMUNICATION

Unlike our other woodpeckers, the Green Woodpecker rarely drums. Any drumming that does take place is rather feeble and unconvincing, typically carried out next to the nest cavity – perhaps serving as a signal to the bird sitting inside. However, the species is known for its loud, laughing call, which also gives rise to the local country name of 'yaffle'.

BTO BIRDFACTS

GREAT SPOTTED WOODPECKER
Dendrocopos major

FAMILY	Picidae
LENGTH	22cm
WINGSPAN	36cm
WEIGHT	85g
HABITAT	Deciduous woodland and wooded suburbs.
FOOD	Mostly insects taken from within dead wood, but also tree seeds, bird eggs and nestling songbirds.
NEST	Hole excavated in a tree.
EGGS	4-6 eggs, white in colour.
EGG SIZE	26.4 × 19.5mm
INCUBATION	14-16 days. 1 brood.
BREEDING SEASON	April to July
POPULATION SIZE	44,000 breeding pairs.
TYPICAL LIFESPAN	Not known
MAX. RECORDED AGE	10 years 9 months

The Great Spotted Woodpecker is the larger of our two black and white woodpeckers, being about the size of a Starling (the Lesser-spotted is about the size of a sparrow). It is also the most widespread and a regular visitor to many garden feeding stations.

ON THE MOVE – JUST!

While all woodpeckers are predominantly sedentary in lifestyle, the Great Spotted Woodpecker does show a greater degree of movement than most. British Great Spotted Woodpeckers don't tend to move far but those from populations living in the northern pine forests of Scandinavia show eruptive movements when the seed crops fail. Such movements bring a small number of birds from Scandinavia to Britain each autumn and periodically there are much larger influxes – the most recent of which was in 1974 when birds arrived in Scotland.

NESTBOX PREDATOR

It surprises many people to discover that Great Spotted Woodpeckers are regular nest predators, breaking into nestboxes to reach the chicks inside. Some individuals enlarge the nest hole to gain access but others drill a fresh hole in the side of the box. Most interestingly, individual Great Spotted Woodpeckers may tap the nestbox first before breaking in. This behaviour is an extension of that used when testing for the presence of beetle larvae in dead wood, the tapping drawing a response from both the beetle larvae and any young chicks, unaware that a response may seal their fate.

IT'S A FACT...

Great Spotted Woodpeckers not only wedge nuts into cracks in bark to ease their opening but are the only woodpecker to fashion anvils for the same purpose.

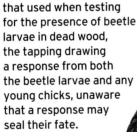

male

A USEFUL TOOL

The Great Spotted Woodpecker population increased throughout the 20th century and the species is now common in suburban parks and gardens. Even so, it is more of a woodland specialist than the Green Woodpecker. The species is absent from higher ground within Scotland and from Ireland and the Isle of Man. Perhaps surprisingly for a bird seemingly adapted to extract invertebrates living inside dead and dying wood, the Great Spotted Woodpecker eats a lot of tree seeds. These are of particular importance during the winter, when the birds favour the seeds of pine and spruce, though they will also take those of oak, beech and hazel. Sometimes the seeds are wedged into the bark of trees and then hammered open with the chisel-like bill. This bill is adapted for getting into dead wood to reach insects but it also allows the woodpecker to 'drum', producing a sound that can be heard over long distances. Drumming is most often heard between March and May. Both sexes drum. A loud, distinctive 'tchick' call may also be heard from time to time.

male

BREEDING BEHAVIOUR

Great Spotted Woodpeckers typically only pair for a single year, otherwise maintaining separate feeding territories outside of the breeding season. Having said this, courtship can begin as early as December, with males drumming from suitable trees, so you may well see a pair together early in the year. An established male will tend to maintain the same territory from one year to the next, often using the same tree for the nest cavity - though not the same hole as used the previous year. Another feature of the early breeding season is the tendency for the woodpeckers to become involved in aerial pursuits, noisily enacted through the still bare tree tops. The breeding season proper does not begin until April, the pair cementing their bond and excavating a nest hole within which the eggs will be laid. Both parents share the responsibility for incubation, the male incubating at night, the female during the day. Once the eggs hatch, the resulting chicks are tended by both parents, with the male continuing to brood them overnight. Young Great Spotted Woodpeckers can be very noisy in the nest, where they are often fed on a diet of moth larvae. When they leave the nest hole during early summer there is an increase in the Garden BirdWatch reporting rate, as adults bring their young to garden feeding stations. Peanuts are a favoured food, nutritionally similar to many tree seeds, and visits to hanging feeders by family parties provide an excellent opportunity to try your identification skills. Later in the year these family parties break up and the young move away.

male

female (left) and juvenile (right)

IDENTIFICATION

Juveniles have a red crown and pale red coloration under the tail. Adult males show a red patch on the back of the head and bright red undertail coverts. Adult females also show this bright red undertail but lack any red on the head. All ages of the Lesser Spotted Woodpecker lack any red under the tail.

STATUS

Rapid increase in breeding population since the early 1990s, hence Green listed.

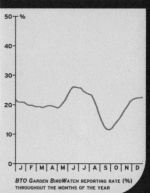

male

male   female

SEASONAL TRENDS

BTO Garden BirdWatch reporting rate (%) throughout the months of the year

BTO BIRDFACTS

FAMILY	Hirundinidae
LENGTH	18cm
WINGSPAN	34cm
WEIGHT	19g
HABITAT	Farmland and open country, often near water.
FOOD	Aerial insects, especially flies.
NEST	Shallow cup of mud pellets, mixed with plant material and lined with feathers. Typically on a small ledge up against a vertical surface.
EGGS	4-5 eggs, white in colour and lightly marked with red-brown spots.
EGG SIZE	20.0 × 13.7mm
INCUBATION	17-19 days. 2 broods.
BREEDING SEASON	March to October
POPULATION SIZE	680,000 pairs.
TYPICAL LIFESPAN	3 years
MAX. RECORDED AGE	11 years 1 month

SWALLOW
Hirundo rustica

In Britain and Ireland Swallows are strongly associated with Man and virtually all nests are built within man-made structures, such as barns, porches and outhouses. The most favoured sites are often those in old farm buildings with livestock present, set within landscapes where large aerial insects are abundant.

MUD GLORIOUS MUD

Swallow nests are made from pellets of mud, collected and mixed with saliva and fibrous material to give the nest rigidity. Many of the nests will last more than one season and both birds from a pair may return to occupy a nest used during a previous year. The male usually arrives before the female and will either begin to construct a new nest or make repairs to an existing one, leaving the female to take over when she arrives. During the course of building the nest, birds may collect more than 1,000 separate mud pellets. Many pairs manage to raise two broods, while some successfully rear three. This is one of the reasons why they are one of the last migrants to leave Britain and Ireland. If suitable nest sites are available then Swallows may nest in small groups, although the individual pair defends the area around its own nest and they cannot really be regarded as being colonial. The newly fledged young will remain around the breeding site, being fed by their parents, for up to six weeks but are often chased away by the male earlier than this. Once they leave the local area these youngsters join large communal roosts, where they will remain until the southward migration begins.

A SINISTER SIDE

In the struggle to see their genes passed on to the next generation certain unpaired males will employ a range of tactics to secure a mate. They often visit the nests of other Swallows in an attempt to associate with a paired female. By presenting himself as a potential suitor, the male is more likely to be accepted as a mate by the female if her own mate dies or the nesting attempt fails and the established pair divorces. Sometimes, rather than simply associate in the hope of a favourable outcome, the male will take things into his own hands. If he finds an unattended nest containing young chicks then he may kill the entire brood, making it more likely that the established pair whose nest has been destroyed will divorce.

Juvenile begging for food.

THE LONG FLIGHT HOME

Most Swallows leave Britain and Ireland during September and October, heading south through Europe and into North Africa, where they cross the Sahara. During November our Swallows finally reach their wintering grounds in South Africa, where they will remain until February. The return journey follows the reverse route and brings the first breeding birds back into southern Britain during March. Poor weather conditions along the route, especially cold and wet weather, can cause serious losses. The first returning birds often face relatively poor conditions on the breeding grounds and may search for prey over water, where flying insects tend to be more abundant. The birds may also associate with livestock, since cattle and horses often disturb flies from the vegetation in which they feed. Changing farming practices, including more intensive stock rearing, modern buildings, pesticides and improved farm hygiene, may exert an impact on breeding populations of Swallows, by reducing the abundance of suitable insect prey. There is certainly evidence of a long-term decline in the abundance of flying insects within parts of Britain. Swallow populations are known to fluctuate from one year to the next, especially at the local level, and this is most probably the result of mortality occurring during migration or on the African wintering grounds. Swallows undertake incredible journeys for such small birds (it is something like 9,500km to South Africa where many winter) and do so year after year.

IT'S A FACT...

Early naturalists wrongly thought that Swallows hibernated through the winter, rather than undertaking the marathon journey south to Africa.

WHO'S A PRETTY BOY THEN?

The long tail feathers of adult Swallows play an important role in mate selection. Research has demonstrated that female Swallows find males with the longest tail feathers the most attractive. These males preferentially acquire mates earlier than males with shorter tail feathers and, importantly, they also enjoy higher breeding success. It is thought that such adornments sported by the males of some species are a true indication of the 'fitness' of a male. The females can then use these features to select the best mates. Young birds lack the long streamers and the rufous colouring is less strongly marked than in the adults. Swallows of all ages lack the white rump that is characteristic of the House Martin. The flight is swift, low and with lots of banking and turning as they search for large flies.

juvenile

male

IT'S A FACT...

Swallows feed on aerial insects but tend to forage at a lower altitude than either Swifts or House Martins.

STATUS

Amber listed due to long-term decline across Europe.

SEASONAL TRENDS

```
50 %
40
30
20
10
0
   J F M A M J J A S O N D
```

BTO GARDEN BIRDWATCH REPORTING RATE (%)
THROUGHOUT THE MONTHS OF THE YEAR

FAMILY
Hirundinidae

LENGTH
12cm

WINGSPAN
28cm

WEIGHT
19g

HABITAT
Open country and towns.

FOOD
Aerial insects, especially small flies and aphids.

NEST
A cup of mud attached to a building under the eaves.

EGGS
4-5 white eggs, very occasionally with fine red spotting.

EGG SIZE
19.4 × 13.4mm

INCUBATION
13-19 days. 2 broods.

BREEDING SEASON
April to September

POPULATION SIZE
Between 250,000 and 500,000 breeding pairs.

TYPICAL LIFESPAN
2 years

MAX. RECORDED AGE
7 years 1 month

IT'S A FACT...

A few birds manage to breed aged up to six years, although they rarely live longer than this.

HOUSE MARTIN
Delichon urbica

The House Martin shows a very close association with Man, building its familiar mud nests under the eaves of houses, outbuildings and other man-made structures (e.g. bridges). These nests are usually placed in colonies, which can contain up to several hundred pairs.

HERE TO BREED

House Martins are summer visitors and the main influx into the country occurs during the second half of April and early May. In common with many other migrants, males tend to arrive back before females and older birds return before young birds. The older birds usually occupy old nests and, while a few running repairs may be needed, they are spared the arduous task of building a nest from scratch. The nests are made from mud and are built from the base up over a period of one to two weeks. Each nest may be made up of over 1,000 beakfuls of mud. During periods of very dry weather, the birds can have difficulty in finding sources of mud. The males are surprisingly aggressive in the defence of their small nest territories but they sometimes lose out to House Sparrows, which will take over nearly completed nests. Most House Martins usually go on to rear a second brood (very few manage three) and are helped on occasion by young from the first brood. This helps to ensure that the young from the second brood receive sufficient food to fledge before the summer ends, while at the same time giving the young from the first brood some experience in rearing chicks.

IT'S A FACT...

You might imagine that House Martins compete for aerial insects with Swallows, Swifts and Sand Martins. To some extent they do, but the different species tend to feed at different heights and on differently sized insects, thus reducing the degree of competition.

Partly constructed nest, built using mouthfuls of mud.

TROUBLED TIMES

Despite its association with Man, the House Martin has proved a difficult species to monitor and we are not completely sure about how it is doing. Reports of the sudden disappearance of long-established colonies can be alarming but these need to be set against the establishment of new colonies and the continued use of others. Data from the BTO's monitoring schemes suggest that there has been a gradual long-term decline in the size of the breeding population, a pattern mirrored in other parts of Europe. More work is clearly needed to understand what is behind this decline but it could be challenging, not least because the wintering areas (and what is happening in them) remain largely unknown. The destruction of House Martin nests by householders dismayed by the mess the birds make may also be a problem. One solution, though, is to erect a tray just beneath the nest onto which the droppings will fall, sparing the windows and ground below.

WILL THEY BE OK?

Each year worried garden birdwatchers seek advice on late broods of House Martins, still in the nest during September or even early October. 'Will these birds be able to migrate south successfully?' Well, the answer appears to be 'it depends'. House Martins are one of the last migrants to depart in the autumn and, since they are aerial feeders, they do not have to fuel up before they go. However, if the weather suddenly deteriorates the all-important aerial insects disappear and the House Martins are caught out. In some years these late broods survive and migrate south, in others they perish.

IT'S A FACT...

In dry summers, House Martins can be helped by creating mud 'puddles' to provide a source of soft mud for nest building.

WHERE DO THEY WINTER?

Even though large numbers of House Martins have been ringed in Britain and Ireland, we still know surprisingly little about where they spend the winter. Results from ringing right across Europe have highlighted that House Martins winter in Africa, south of the Sahara, but we do not know the exact wintering locations of birds from the different breeding populations. It has been suggested that the birds may remain on the wing, feeding above the equatorial forest canopy, where they do not come into contact with Man. This would explain the lack of ringing recoveries.

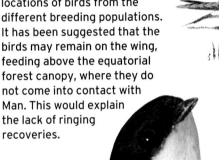

The House Martin has shiny blue-black and white plumage with a white rump. This white rump is distinctive and is absent from our other Swallows and martins. The species has a short and moderately forked tail, but lacks the long tail streamers carried by the Swallow.

STATUS

Amber listed because of the steady long-term decline.

SEASONAL TRENDS

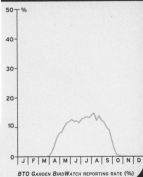

BTO *Garden BirdWatch* reporting rate (%) throughout the months of the year

PIED WAGTAIL
Motacilla alba

With its black and white plumage and constantly wagging tail, the Pied Wagtail is familiar to most garden birdwatchers. The Pied Wagtail is actually the British and Irish race of the White Wagtail, which breeds elsewhere in Europe (and sporadically here).

FAMILY
Motacillidae

LENGTH
18cm

WINGSPAN
28cm

WEIGHT
21g

HABITAT
Open country, often near water, and urban centres.

FOOD
Small invertebrates.

NEST
A cup of grass, roots and twigs, lined with wool, hair and feathers.

EGGS
5-6 eggs, white or grey-blue and covered with fine grey-brown freckles.

EGG SIZE
20.6 × 15.3mm

INCUBATION
13 days. 2 broods.

BREEDING SEASON
April to August

POPULATION SIZE
255,000 to 330,000 breeding pairs.

TYPICAL LIFESPAN
2 years

MAX. RECORDED AGE
11 years 3 months

FINDING FOOD

Pied Wagtails utilise a wide range of habitats and are even found nesting in the middle of our largest urban centres. Wagtails specialise in feeding on insects, especially small flies like midges, often taken at the water's edge. During the late autumn, they may also be seen at garden feeding stations taking seeds and even bread; foods that will be additional to the insects upon which they are likely to be concentrating. During the winter months, things become more difficult because insects become less abundant. It has been discovered that Pied Wagtails may spend more than 90% of the short winter daylight hours feeding, taking a small prey item every three to four seconds. Even this concentrated effort may not be enough to meet the energy requirements of the bird and this may be why northern populations tend to migrate south for the winter, where temperatures may be higher and prey more abundant. Adult males establish feeding territories in the winter and, if sufficient food is available within them, may tolerate the presence of females or immatures. These 'satellite' birds help to defend the territory against other birds, but are soon chased off if food availability declines. Other birds remain in feeding flocks that are more mobile. Pied Wagtails can often be seen walking about, occasionally pecking some prey item from the ground but on other occasions making a darting run to grab something that has caught their eye. They may also 'flycatch' from a suitable perch.

RACES

Britain and Ireland together hold almost the entire population of the distinctive dark-backed race of the White Wagtail, the bird known to us as the Pied Wagtail. A different race, the true White Wagtail, occurs elsewhere in Europe. While some of our Pied Wagtails are resident, others migrate south in late autumn, with those from Scotland wintering in southern England or across the Channel in France and Spain. Some of those from southern Britain also migrate south, again wintering in France and Spain, with a few reaching North Africa. White Wagtails are completely migratory, moving to the Mediterranean Basin and extending south from there to the tropical regions of Africa. Given that this species is dependent upon invertebrates, it is easy to see why many are forced to move south to warmer climes during the winter. In southern Britain, look for migrant White Wagtails mainly on the coasts - or sometimes beside inland lakes and gravel pits - in spring and autumn. Their migratory instincts are strong and they seldom linger.

Pied Wagtail adult male

White Wagtail adult male

IDENTIFICATION

The distinctive bounding flight and sharp 'chisick' flight call, coupled with bold black and white plumage should ensure that the Pied Wagtail is instantly recognisable. Our other two wagtails (the Grey Wagtail and Yellow Wagtail) are not normally seen visiting gardens. While the brilliantly coloured Yellow Wagtail is an extremely rare garden visitor - preferring lowland wet meadows and farmland - some lucky garden birdwatchers are visited by Grey Wagtails during the winter months. It is at this time of the year that Grey Wagtails move to lowland stretches of river. At the same time, our breeding Yellow Wagtails will be on their wintering grounds in Africa.

NESTING

The Pied Wagtail's nest is a cup of twigs and grass stems, lined with wool, hair and feathers, and placed in a wide variety of recesses, including the walls of old houses and outbuildings. The species can sometimes be tempted to nest in open-fronted nestboxes of the type used for Robins. Large gardens, close to running water, are favoured.

BIRDS TOGETHER

Another feature of the winter ecology of this species is the large communal roosts that can gather at favoured sites. Hundreds of individuals may congregate to roost in reedbeds, around buildings such as hospitals and supermarkets or in large commercial greenhouses. Not only do such roosting sites provide much-needed warmth, they can also act as information centres. Birds that have not fed that well during the day are able to spot individuals that have been more successful; the following morning they will follow these individuals out of the roost and, with luck, have a better day. One final benefit of roosting communally is safety in numbers. While the presence of many small birds at a roost might attract predators, the individual's chances of being taken by a predator remain lower than they would be if the bird were alone.

juvenile

With a tail that is often pumped excitedly, this long-legged insect eater will make sudden short runs to grab suitable prey.

Yellow Wagtail

Grey Wagtail

Pied Wagtail adult female

Pied Wagtail adult male

STATUS

Suggestion of slight decline in numbers over recent years but longer-term trend is favourable, hence Green listing.

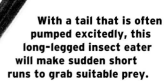

SEASONAL TRENDS

```
20 %

16

12

8

4

0
   J F M A M J J A S O N D
```

BTO GARDEN BIRDWATCH REPORTING RATE (%) THROUGHOUT THE MONTHS OF THE YEAR

IT'S A FACT...

The Pied Wagtail has a wide distribution, being found from Greenland in the north and Morocco in the south, across Europe, Asia and into western Alaska.

BIRDFACTS

adult

WAXWING
Bombycilla garrulus

FAMILY
Bombycillidae

LENGTH
18cm

WINGSPAN
34cm

WEIGHT
63g

HABITAT
Winter visitor to hedgerow fruits and garden shrubs.

FOOD
Berries in winter, insects in summer.

NEST
Does not breed in Britain but nests, in native range, are often high in a tree.

EGGS
5 eggs, grey-blue to buff in colour, lightly spotted dark grey.

EGG SIZE
24.6 × 17.2mm

INCUBATION
13-14 days. 1 brood.

BREEDING SEASON
May to July

POPULATION SIZE
Number of winter visitors to Britain can vary from a few dozen individuals up to 12,000. Does not breed here.

TYPICAL LIFESPAN
Unknown

MAX. RECORDED AGE
2 years 11 months

There can be few more exciting garden visitors than the Waxwing, with its striking plumage and delightful feeding habits, arriving in small flocks to feed on berries, which it plucks with acrobatic ease from ornamental shrubs and bushes across parks, gardens and supermarket car parks.

A WINTER VISITOR

The Waxwing is a winter visitor to Britain and Ireland, breeding far from here in the great belt of boreal forest that stretches from Scandinavia, across Russia through to the Pacific and then again across parts of western North America. That individuals from these distant breeding populations sometimes reach Britain is all down to the availability of the bird's favoured winter food – the berries of Rowan *Sorbus aucuparia*. Waxwing populations typically breed at low densities but in favourable conditions their populations are swelled when an abundance of rowan berries allows a generous winter survival of young from the previous year. Typically, the following winter will see a crash in the abundance of these berries, the trees having fruited so well the previous year producing a much smaller crop as they struggle with reduced resources. This triggers the Waxwings to move away from their breeding grounds in search of berries elsewhere and such movements can see the birds cover exceptionally long distances. These movements do not happen every year but, when they do, they can bring thousands of Waxwings to northern and eastern Britain, with much smaller numbers filtering across to the extreme southwest of the country.

The earliest arrivals tend to be reported from eastern Scotland and the Northern Isles. If the arrival begins early in the autumn then it can signal that a much bigger movement will follow later in the year. In most years there are at least a few Waxwings wintering here but in peak years numbers can reach 12,000 individuals or more.

male

female

FAVOURED FRUITS

Although our native rowan, *Sorbus aucuparia*, is favoured by visiting Waxwings, the berries of other native and non-native *Sorbus* species are also taken, including those with white, yellow or even pink berries. It is not just the rowans and their relatives that are exploited; Waxwings will also turn to the berries of hawthorn and Cotoneaster, the latter proving to be important later into the winter. Some individuals can also be seen feeding on insects, flowers and tree buds. That insects are taken is hardly unsurprising, as Waxwings feed heavily on mosquitoes and midges during the breeding season. Watch a flock of feeding Waxwings and you will notice that the berries are invariably taken from the tree, rather than the ground, and are plucked with a slight stooping motion before being swallowed with a quick toss of the head. For such a large bird (they are about the size of a Starling) they are surprisingly agile, and may be seen clinging to the underside of the branch and eating the berry from below.

MOVEMENTS WITHIN BRITAIN

With Waxwings arriving first to eastern parts of Scotland, there is then a tendency for the birds to disperse further south through the winter. The speed of this dispersal is dependent upon the berry crop in those areas where the birds first arrive. If the crop is poor, and quickly depleted, the birds will move on fairly rapidly. If sufficient berries are present then they may remain in the north and east of the country for some time.

Female, contemplating a meal of dried, shrivelled Rowan fruit.

ATTRACTING WAXWINGS

If you want to attract Waxwings to your garden then you need to plant plenty of berry-producing plants. Many of the varieties of *Sorbus aucuparia* are suitable as are some of the Cotoneasters, including *Cotoneaster horizontalis*. If Waxwings are in the area, then you can even entice them onto halves of apple tied to a branch.

Feasting on ripe Rowan fruits.

IT'S A FACT...

One reason why these unusual winter visitors turn up in gardens and supermarket car parks is that these are exactly the sorts of places where berry-producing shrubs and tress are planted.

IDENTIFICATION

At first glance, a flock of Waxwings silhouetted against the sky can resemble a flock of Starlings. However, look more closely and you will notice the characteristic crest and the soft peach-brown tones of the plumage. The most striking features are the narrow black eye mask and chin, the bright yellow tip to the tail and the long waxy red appendages to the white secondary wing feathers. It is this last feature that gives the Waxwing its name.

STATUS

Not a species of concern, Green listed.

Once evey decade or so, large numbers of Waxwings arrive all at once, and fortunate birdwatchers have a superb opportunity to observe many birds together.

SEASONAL TRENDS

MORE DATA IS NEEDED TO COMPILE AN ACCURATE CHART FOR THIS SPECIES - *PLEASE* KEEP SENDING IN YOUR RECORDS

J F M A M J J A S O N D

BTO GARDEN BIRDWATCH REPORTING RATE (%) THROUGHOUT THE MONTHS OF THE YEAR

BTO BIRDFACTS

WREN
Troglodytes troglodytes

Tiny, restless and pugnacious, the Wren is more often heard than seen, skulking in the undergrowth to feed on insects and spiders and to deliver its surprisingly loud song. Although it is one of our most adaptable and widespread birds, Wren numbers can fall dramatically following particularly cold winters.

THE NESTING HABIT

Wrens like to nest in the understorey of broad-leaved woodland, especially the damper areas where the vegetation is particularly lush. Other habitats, including gardens and orchards, are increasingly used when numbers are buoyant. Males establish their breeding territories in early spring and are highly aggressive towards each other. Each male will court any female that enters his territory, tempting her to use one of the nests that he has built. Wrens in southern Britain will build five or six nests, the female visiting each in turn to determine which, if any, she wishes to use. Once selected, the female lines the nest with feathers. Males nesting further north build fewer nests and are more attentive partners than those in the south: tougher environmental conditions (lower food availability and a shorter breeding season) mean they have to work harder to rear their young. Those in the more comfortable conditions of southern Britain can indulge in establishing relationships with multiple females, at the same time investing less time and energy in each individual nesting attempt. Similarly, the species tends to be single-brooded in the extreme north of Britain, managing two broods further south.

FAMILY	Troglodytidae
LENGTH	10cm
WINGSPAN	15cm
WEIGHT	7–12g
HABITAT	Woodland and other habitats with undergrowth.
FOOD	Mostly insects (especially small beetles) and spiders. May take small seeds.
NEST	Domed structure of moss and grass, usually in a crevice or hole.
EGGS	5–6 white eggs, sometimes with brown speckling at broad end.
EGG SIZE	16 × 13mm
INCUBATION	16–18 days. 1, sometimes 2 broods.
BREEDING SEASON	March to August
POPULATION SIZE	8 million breeding territories in summer.
TYPICAL LIFESPAN	2 years
MAX. RECORDED AGE	6 years 8 months

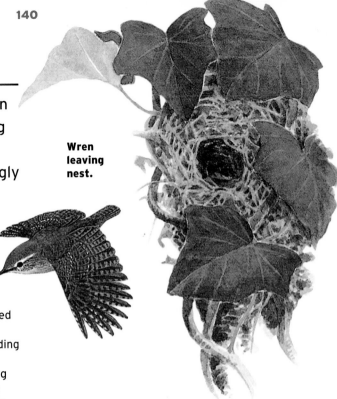

Wren leaving nest.

IT'S A FACT...

Wrens feed on insects and spiders, occasionally adding small seeds to their diet. Their small size and agility mean that Wrens can access crevices out of the reach of other birds.

ISLAND FORMS

Not all of our Wrens are the same and four island races exist that are distinctly different from those seen across mainland Britain and Ireland. Shetland, Fair Isle, St Kilda and the Outer Hebrides each have their own form, different in colour, song and, to some extent, behaviour. Separated from the mainland, island populations have diverged in character and are usually darker in colour and more heavily barred than the Wrens you are likely to see visiting your garden.

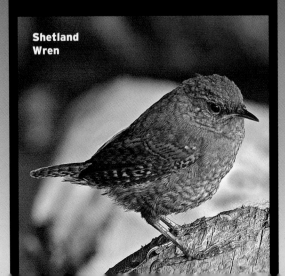

Shetland Wren

STUNNING SONGSTER

For such a small bird, the Wren has a very loud song, something which poses the question of how such volume is produced. Unlike humans, the larynx of birds lacks any vocal chords and so has little or no role in song production. Instead, Wrens and other birds have an organ known as a syrinx, located just above where the windpipe forks into the two lungs. The syrinx has several components, including a resonating chamber (the tympanum), a series of vibrating membranes and controlling musculature. The syrinx is far more efficient at producing sound than our own larynx and its associated vocal chords. While our vocal chords can only manage to utilise just 2% of the air passing over them, the syrinx is able to vibrate just about all of the air coming out of the lungs. In addition, the syrinx can produce two different sounds at the same time, something that helps to explain the complexity of sounds produced by a singing bird.

Foraging Wren.

Singing Wren.

THE WREN IN FOLKLORE

The Wren features quite heavily in folklore, with a tradition of 'hunting the Wren' still played out in parts of Ireland. The central focus of this particular custom revolves around groups of young boys going out into the countryside to capture or kill a Wren, which is then paraded around the village. The boys demand rewards, in the form of cakes or money, for the service they have provided their community. Quite why the Wren is the subject of such victimisation seems to vary from place to place. In some areas the Wren was targeted for having betrayed the Irish to their enemies, in others because it was thought to be a fairy who lured young men to a watery grave.

Singing Wren.

WINTER TERRITORIES

After the breeding season ends, Wrens begin to establish their winter territories, with those individuals unable to secure a territory moving into other habitats (like reedbeds) not normally used for breeding. Many birds will remain within these habitats until March. Holding a territory during winter has several benefits for a small bird, not least providing familiarity with where food and roosting sites can be found. Competition for territories can be fierce, reflecting the fact that birds that secure winter territories are more likely to survive the winter than those without.

IT'S A FACT...

Being small has its disadvantages; Wrens lose body heat quickly and are unable to build up stores of fat for the hard times of winter.

ROOSTING

Wrens may roost together in winter to reduce heat loss during the long winter nights. Most roosts contain just a handful of individuals but up to 60 have been reported roosting together, Nest boxes and roosting pouches are favoured, the birds arriving shortly after dusk and departing before first light.

Wren in snow.

STATUS

Widespread and stable, hence Green listing.

SEASONAL TRENDS

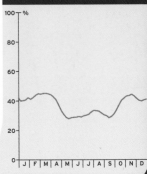

BTO GARDEN BIRDWATCH REPORTING RATE (%) THROUGHOUT THE MONTHS OF THE YEAR

BTO
BIRDFACTS

FAMILY	Prunellidae
LENGTH	14cm
WINGSPAN	20cm
WEIGHT	21g
HABITAT	Scrubby undergrowth, woodland and gardens.
FOOD	Mostly insects, spiders and small seeds taken from the ground under cover.
NEST	A cup of twigs, leaves and roots, lined with moss, hair and wool, usually well hidden in thick cover up to 3m off the ground.
EGGS	4-5 bright blue eggs, very occasionally marked with a few reddish spots.
EGG SIZE	19.9 × 14.7mm
INCUBATION	14-15 days. 2-3 broods.
BREEDING SEASON	March to August
POPULATION SIZE	2.1 million breeding territories.
TYPICAL LIFESPAN	2 years
MAX. RECORDED AGE	11 years 3 months

DUNNOCK
Prunella modularis

This unobtrusive and rather plain looking bird is sometimes called a 'Hedge Sparrow' but, while it does favour hedgerows and other low cover, it is not a sparrow at all. Instead, it belongs to a group of birds known as accentors, which specialise in feeding on insects and small seeds, hence the thin Robin-like bill.

IT'S A FACT...
Male Dunnocks can sometimes be seen defending their territories, flicking their wings in an agitated fashion.

SOCIAL SYSTEMS

Perhaps the most striking aspect of Dunnock behaviour is its complex social system. While many Dunnocks maintain monogamous relationships, others indulge in more complex arrangements, made possible because males and females maintain their own, largely independent, breeding territories. Those of the males tend to be larger in size and so a male will often have access to two or more females, whose own territories either fall within or overlap his own. Where this happens it would normally lead to polygyny (a breeding system where a male regularly mates with two or more females). However, this is not an unequal system because the females will often mate with more than one male (termed polyandry). The Dunnock's social system is complicated further by the fact that some male territories may be shared by two males, one of whom (the 'alpha' male) is dominant over the other (the 'beta' male). This is an uneasy arrangement, the beta male maintaining his position within the alpha male's territory through sheer persistence, and the alpha male spending a lot of time guarding his female(s) from the advances of the beta male (or other intruding males). Curiously, both the alpha and beta males will work together to proclaim territory ownership and to drive off intruders.

SEX IN THE SHRUBBERY

A particularly fascinating piece of behaviour results from the Dunnock's complex social arrangements. Prior to copulation, a female Dunnock will crouch low in front of the male, fluffing out her feathers, lifting her tail and quivering. The male responds to this by positioning himself behind his mate and then, while hopping from side to side, he will peck at the female's cloaca. As the pecking continues, so the cloaca becomes enlarged, occasionally making strong pumping movements, which eject some of the sperm from previous matings. This behaviour is the male's way of attempting to increase his contribution to the paternity of any resulting offspring. This may seem a little harsh on the female but she also plays the odds by deliberately soliciting matings with other males (especially the beta male). By doing so she hopes to secure sufficient attention from more than one male during the important period of chick rearing. With hungry mouths to feed, the more males she can persuade to provide food, the greater her chances of successfully rearing her brood.

A CUCKOO IN THE NEST

The Dunnock is one of the main hosts for the Cuckoo, a species renowned for laying its eggs in the nests of other birds – a behaviour known as brood parasitism. A female Cuckoo will remove one of the Dunnock's eggs before laying one of her own. Cuckoos target a number of species, including Dunnock, Robin, Meadow Pipit, Reed Warbler and Pied Wagtail. The eggs of the different host species vary in colour and pattern and, since an unusual looking egg is likely to be rejected by the host, there is a selection pressure on the female Cuckoo to produce eggs that closely resemble those of the host. This is why individual female Cuckoos tend to specialise on one particular host. Cuckoos that target Reed Warbler and Meadow Pipit produce eggs that closely match those of their host. However, in the case of the Dunnock, the Cuckoo eggs look very different from the blue eggs produced by the Dunnock and you might expect them to be rejected by the parent birds. Interestingly, Dunnocks rarely reject Cuckoo eggs and this suggests that they are not particularly discerning parents. Since it is in the Dunnock's interest to spot the Cuckoo egg, it seems likely that the Dunnock is a relatively recent host and has not yet shown an evolutionary response to this act of nest parasitism. Ending up with a Cuckoo in your nest may seem disastrous but, since a pair of Dunnocks can rear two or even three broods in a season, they could still have the opportunity to rear chicks of their own later in the year.

Dunnock feeding young Cuckoo.

juvenile

FINDING FOOD

Dunnocks are usually to be seen under bushes or below the bird table, foraging with mouse-like movements as they search for insects and small seeds. They rarely venture onto hanging feeders but will visit raised bird tables. During the winter months they take more small seeds and fewer insects and spiders. Dunnocks also take peanut fragments, finely grated cheese, bread-crumbs and niger seed. They have also been noted taking fat, small pieces of meat and various berries. Although usually solitary in winter, they may gather together to exploit particularly rich food sources.

IN DECLINE !

There has been a steady decline in the size of the British and Irish Dunnock population over recent decades, a pattern that is mirrored in results from the BTO's Garden Bird Feeding Survey, and this has led to the species being flagged as of conservation concern. The exact cause of the decline is unknown but a similar pattern has been seen in other European countries. However, the population has fared somewhat better over the last few years, which is an encouraging sign for the future.

In winter Dunnocks will sometimes forage underneath bird tables.

STATUS

Amber listed because of long-term decline in numbers.

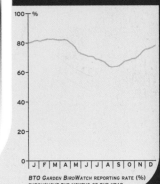

SEASONAL TRENDS

100 %
80
60
40
20
0

J F M A M J J A S O N D

BTO GARDEN BIRDWATCH REPORTING RATE (%) THROUGHOUT THE MONTHS OF THE YEAR

BTO BIRDFACTS

ROBIN
Erithacus rubecula

FAMILY	Turdidae
LENGTH	14cm
WINGSPAN	21cm
WEIGHT	18g
HABITAT	Woodland edge, scrub and gardens.
FOOD	Insects, especially beetles, spiders, together with fruit and seeds.
NEST	A bulky structure with leaves for the base on which is woven a cup of moss, grass and more leaves. The cup is lined with finer material, including hair and, occasionally, feathers.
EGGS	4-5 eggs, white or faintly bluish white with sandy red freckles and small blotches, which may sometimes make the egg appear reddish.
EGG SIZE	19.9 × 15.4mm
INCUBATION	14-16 days. 2-3 broods on average.
BREEDING SEASON	March to July
POPULATION SIZE	5.5 million pairs.
TYPICAL LIFESPAN	2 years
MAX. RECORDED AGE	8 years 4 months

Voted the Nation's favourite bird on more than one occasion, the Robin is also one of the most widely distributed of our birds. Because of their approachability and choice of habitats, Robins have been the subject of many studies, including that carried out by David Lack and famously published in 'The Life of the Robin'.

CATHOLIC AND CONFIDING

Robins are catholic in their choice of food items, although most of the diet consists of invertebrates taken from the surface of the ground. Fruit is also important, especially in the autumn and winter. They have been recorded feeding at hanging peanut feeders, and even hovering at hanging fat blocks, but they are happier taking mealworms, peanut cake, fat and finely grated cheese from bird tables or off the ground. Their confiding nature is further demonstrated by the way in which they will often associate with a gardener digging over the soil. This behaviour is probably an extension of the habit that Robins have of taking insects from ground disturbed by foraging moles. Robins usually use one of two main strategies for finding food. Individuals will sometimes sit on a low branch, scanning the ground beneath them. Every now and then they will spot a potential prey item, fly down to grab it, and then return to the perch. The other strategy involves a bit more legwork, with the individual hopping across the ground actively looking for food.

Juvenile birds, recognised by their spotted plumage, are dependent on the parents for food for a short while after they fledge.

YEAR ROUND SONG

Most male Robins defend territories throughout the year, which is why their plaintive song can be heard in every month. The territory structure adopted by Robins is all-important and only breaks down during periods of severe winter weather or when the birds are skulking away during their annual moult. Many females also establish winter territories of their own, often close to where they will breed the following year. Other females appear to move farther away from their breeding territories, sometimes setting up winter territories in habitats that could not support a nesting pair but which can support an individual Robin through the winter. Towards the end of the winter, these females return to their original territory and partner from the previous year.

MOVING AROUND EUROPE

A small number of our Robins appear to spend the winter on the Continent, joining other Robins that pass through southeastern Britain in the autumn en route south. These passage migrants come from Scandinavia, Germany and the Baltic states and are heading towards wintering grounds in Spain, Portugal and North Africa. Studies have shown that many return to the same winter territory each year, suggesting that these birds have distinct summer and winter territories, many hundreds of kilometres apart, to which they faithfully return each year.

ABLE TO BOUNCE BACK

As BTO Garden BirdWatch results demonstrate, Robin populations have been remarkably stable in recent years. Although they can be knocked back by very bad winters, the ability to produce three or even five broods in a year means that populations recover quickly from overwinter losses. In mild winters, breeding may begin in January and pairs may overlap nesting attempts in order to squeeze in several broods.

WHO'S THE ROBIN IN CHARGE?

Robins seem to take territorial behaviour to an extreme, sometimes attacking other species by mistake. The Robin's red breast is displayed prominently during territorial disputes. Initially, two birds disputing ownership of a territory will sing at each other, the individual owning the territory attempting to take a perch above the intruder. This enables the owner to show off his red breast to maximum effect by adopting a horizontal position and, in most cases, this is usually enough to settle the dispute. If the territory owner finds himself below the intruder then he will adopt a different posture, throwing his head back. Again this emphasises the red breast. Sometimes this display does not do the trick and the dispute escalates to an all-out attack. Such fights can be ferocious and injury or even death may result. Given the risks involved, it is hardly surprising that the birds first attempt to settle the dispute through display rather than all-out aggression.

THE NEST

Robins will take readily to open-fronted nestboxes, so long as these are positioned among some thick cover, such as ivy growing on a wall, where they are afforded some protection from the unwelcome attentions of predatory cats. They will also use garden sheds and greenhouses where considerate owners leave the door ajar.

Robins are colourful even in winter.

STATUS

Population largely stable, hence Green listing.

SEASONAL TRENDS

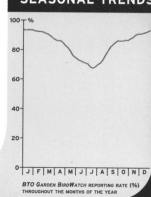

BTO *Garden BirdWatch* REPORTING RATE (%) THROUGHOUT THE MONTHS OF THE YEAR

BTO BIRDFACTS

BLACKBIRD
Turdus merula

female

FAMILY	
Turdidae	

LENGTH
24cm

WINGSPAN
36cm

WEIGHT
100g

HABITAT
Woodland, scrub, gardens and parks.

FOOD
Mainly earthworms and insects. Fruit and berries taken in the autumn and winter.

NEST
Substantial cup of grass and small twigs, positioned in fork of bush or small tree, plastered inside with mud and lined with grass.

EGGS
3-4 eggs, pale greenish-blue and mottled with reddish brown.

EGG SIZE
29.4 × 21.7mm

INCUBATION
13-14 days. 2-3 broods typical but up to 5 in a season.

BREEDING SEASON
March to September

POPULATION SIZE
4.6 million breeding pairs.

TYPICAL LIFESPAN
3 years

MAX. RECORDED AGE
14 years 2 months

Originally a species of high forest, British Blackbirds have spread into human habitats. The transition began in the 19th century and was seemingly complete by the early part of the 20th century. This ability to occupy a range of habitats, especially man-made ones, reflects the adaptable nature of this familiar thrush.

LIVING ALONGSIDE MAN

As much as a third of the British Blackbird population may now live alongside Man, occurring at densities of around 100 birds per km^2, which highlights the importance of urban parks and gardens. Recent research has revealed that Blackbirds nesting in towns and villages are more productive than those in woodland, with the rate of nest predation in gardens (50%) being considerably lower than that in woodland (80%). Woodland populations have to contend with crows, Magpies, Jays, Grey Squirrels and Weasels, all important nest predators, while in gardens only the domestic cat and Magpie are real threats. The most serious problem in gardens, during the nesting period at least, is that of starvation of the nestlings during dry periods when earthworms become difficult to extract from parched lawns. Even allowing for this, and the slightly higher mortality levels that fledged young suffer, suburban populations of Blackbirds are thought to act as a source from which birds can enter the less productive woodland population.

THE GARDEN YEAR

Much of our understanding of Blackbird ecology and behaviour comes from studies carried out in urban and suburban parks and gardens, where traditional breeding and feeding sites are occupied year after year by well-established and dominant birds. Food is usually readily available throughout the year, partly a consequence of the catholic diet of the species, and birds can maintain compact and tightly packed territories. The BTO Garden BirdWatch reporting rate shows a clear seasonal cycle, with a pronounced autumn trough. This trough corresponds to the period when Blackbirds undergo their annual moult, a time when they become somewhat shy and retiring. It is also a time when birds may leave gardens, to feed elsewhere on trees and shrubs packed with autumn fruits.

Earthworms are important in the diet of Blackbirds.

A singing male Blackbird.

Male at the nest.

IT'S A FACT...

Blackbirds will sometimes sing at night, particularly in the presence of street lighting. Along with Robin, another common nocturnal songster, these birds may be mistaken for Nightingales.

BLACK OR BROWN?

Most Blackbirds are brown rather than black; it is only the adult males that are black, females and young males are brown. As a young male moults through into his adult plumage you will see that although he may have black body feathers, his wings remain dark brown. These will only be replaced when he undergoes his first wing moult, a year after moulting his juvenile body feathers. Very young birds are a rich speckled brown and easy to spot as they trail around the garden demanding food from their parents.

male

juvenile

WHITEBIRD?

Some Blackbirds, particularly older males, may develop partly white plumage. In certain individuals the extent of the white gradually increases from year to year - after the moult. This seems to be partly genetically controlled but may also result from damage to the feather shafts.

This pure white Blackbird is at the far end of the leucistic spectrum.

juvenile

IT'S A FACT...

Blackbirds are sometimes observed feeding on tadpoles and newts, taken from the shallows of a garden pond.

STATUS

There has been a recent upturn in fortunes following a period of longer term decline, but Green listed.

ATTRACTING BLACKBIRDS

By supplying food it is easy to attract Blackbirds to your garden. Many are happy to eat kitchen scraps or sultanas (a favourite). Many Garden BirdWatchers prefer to soak the sultanas before putting them out. Blackbirds feel most secure when feeding on the ground, taking food deliberately placed there or spilt from bird tables or hanging feeders. They may come to a raised table but it is an exceptional Blackbird that can manage to use a hanging feeder. Windfall apples are a favoured food in winter and spoiled fruit can often be bought cheaply to attract flocks of Blackbirds.

ON THE MOVE

It has been estimated from bird ringing data that at least 12% of the Blackbirds present in Britain and Ireland during the winter are immigrants from elsewhere in Europe. Our own breeding birds are resident although, following a run of cold winters earlier last century, some breeding populations from northern England and Scotland undertook regular movements to Ireland. Winter immigrants have been shown to originate from countries including Finland, Sweden, Denmark and the Netherlands. Some of these birds pass through eastern England during October to winter further south in France, Spain and Portugal. Others arrive slightly later, to forage alongside our resident birds in orchards and gardens on windfall apples and kitchen scraps. Although it is not possible to identify for certain which are residents and which are immigrants in the winter, by March, when resident males acquire their breeding plumage, the two groups can be told apart. Our resident breeding males will have orange-yellow bills and eye rings, while the immigrant males won't get theirs until they return to their European breeding grounds. Thomas Hardy, who described the bill as being 'crocus coloured', best captured the warmth of the yellow tones.

SEASONAL TRENDS

BTO GARDEN BIRDWATCH REPORTING RATE (%) THROUGHOUT THE MONTHS OF THE YEAR

BTO BIRDFACTS

FIELDFARE
Turdus pilaris

The Fieldfare is the larger of our two winter-visiting thrushes. It is bold in nature and striking in colour but is only really likely to be seen in gardens during the coldest winter weather, a time of year when food may be unavailable in the wider countryside.

ORIGINS
Results from the BTO Garden BirdWatch show that Fieldfares usually arrive in gardens from mid-September (Scotland) to mid-October (southern England). The timing of the main arrival depends on when the birds leave Scandinavia, from where our wintering birds originate. The birds tend to remain on their

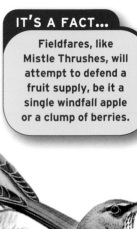

Rowan berries

breeding grounds as long as there is food available, especially Rowan berries. The larger the berry crop, the longer they can remain. Fieldfares are gregarious birds and form large flocks, sometimes in the company of other thrushes or Starlings. These flocks usually feed in open fields, taking insects and earthworms. It is during the hardest weather, when the ground is frozen, that they move into orchards and gardens to feed on windfall apples and berries. This reliance on fruit makes Fieldfares less able to withstand severe weather conditions than some of the other thrushes, which are able to turn to alternative types of food when things get tough. As a result, the winter flocks of Fieldfares tend to be nomadic in nature and there is a general southwards and westwards movement within Britain as the winter progresses and berry supplies run low.

Fieldfares wintering in Britain may spend other winters elsewhere in Europe. Information gathered from the BTO Ringing Scheme has shown how some of our wintering Fieldfares spend subsequent winters on the Rhône and Po floodplains. Within many of these continental countries, Fieldfares may be hunted and trapped in large numbers between September and February, though the amount of hunting that goes on has fallen in recent years.

Flock feeding on windfall apples.

Adult in winter.

IT'S A FACT...
Fieldfares, like Mistle Thrushes, will attempt to defend a fruit supply, be it a single windfall apple or a clump of berries.

FAMILY	Turdidae
LENGTH	26cm
WINGSPAN	40cm
WEIGHT	100g
HABITAT	Woodland and open farmland.
FOOD	Invertebrates, fruit taken all year but especially important in winter.
NEST	A bulky structure of twigs and grass, lined with fine grasses and roots, with a woven ring of grass glued together with mud. Nest placed in a tree, either against trunk or in the fork of a branch.
EGGS	5-6 eggs, glossy pale blue, marked with red-brown speckles.
EGG SIZE	29.1 × 21.3mm
INCUBATION	11-14 days. 1-2 broods.
BREEDING SEASON	May to July
POPULATION SIZE	A handful of pairs may breed in Britain annually.
TYPICAL LIFESPAN	Not known
MAX. RECORDED AGE	11 years 9 months

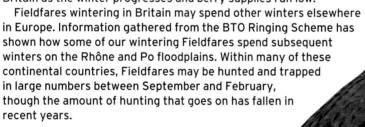

A COLONIAL NATURE

Many Fieldfares nest together in colonies, a habit which is somewhat unusual within the thrush family. It is thought that they do this as a means of protection against predators. While many birds seek to avoid the unwanted attentions of predators by nesting unobtrusively and singly, the large bulky nest of the Fieldfare is difficult to hide. By nesting together the Fieldfares work collectively to drive away any nest predators (like Stoats) that are spotted. The Fieldfares are so determined in the defence of their nests that some predators actively seek out areas away from Fieldfare colonies just to escape the constant harassment. Fieldfares also come together in the winter months and it is rare to see a lone Fieldfare. Not only do flocks of Fieldfares gather to feed but they also roost communally, favouring conifers or thick scrub.

A BRITISH BREEDER?

Small numbers of Fieldfares have bred in Britain over the years, the first documented record coming from Orkney in 1967. Since that time other nesting attempts have been recorded from elsewhere in Scotland and England, possibly reflecting the continued southwest expansion in the European breeding range.

NAMES

The Fieldfare's name varies across the different countries in which it occurs, often reflecting different aspects of its behaviour that have attracted attention from the local inhabitants. The English name comes from the Anglo Saxon word 'feldware', which literally translates as 'traveller of the fields'. This hints at the tendency for winter flocks to feed in open fields, moving around in response to the availability of food. The Swedish name 'Björktrast' refers to its preference for birch woodland, translating as 'birch thrush', a pattern also seen in the old German name for the species. The modern German name of 'Wacholderdrossel' refers to the bird's diet, meaning 'Juniper thrush'. Other names refer to its size (Spain, 'Zorzal royal' - royal thrush), its voice (Denmark 'Sjagger') or its colour (Norway, 'Gråtrost' - grey thrush).

IDENTIFICATION

When seen well, this handsome thrush is relatively easy to identify. The overall appearance is of a large thrush (Blackbird sized but appearing bigger and more upright) with a blue-grey head, paler grey rump, red-brown wings, black tail, heavy spotting below and a yellow ochre wash across the chest. If you get a good view you should be able to see how the spots on the upper breast slowly turn into dark chevrons on the lower breast and flanks of the bird. The call is distinctive and a dry 'chook-chook-chook-chook' is often uttered as the bird takes to flight.

A Fieldfare's pale grey rump is diagnostic in flight.

The chestnut back and yellow-flushed breast make for colourful plumage.

STATUS

Currently Amber listed.

SEASONAL TRENDS

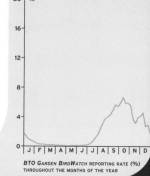

A Fieldfare's clean, white underparts give it a bright white-looking appearance.

BTO GARDEN BIRDWATCH REPORTING RATE (%) THROUGHOUT THE MONTHS OF THE YEAR

BTO BIRDFACTS

REDWING
Turdus iliacus

FAMILY
Turdidae

LENGTH
21cm

WINGSPAN
34cm

WEIGHT
63g

HABITAT
Open farmland, woodland.

FOOD
Invertebrates throughout the year, with increasing amounts of fruit and berries through the autumn and winter.

NEST
Cup of grass, twigs and moss, plastered inside with mud and lined with fine grass. Nest is usually in thick cover of bush or shrub, though may be on the ground.

EGGS
4–5 eggs, pale- to greenish-blue in colour, covered with red-brown speckles.

EGG SIZE
25.9 × 19.1mm

INCUBATION
11–15 days. 2 broods.

BREEDING SEASON
May to August

POPULATION SIZE
Has bred here on occasion but primarily a winter visitor. Numbers arriving may vary greatly from one year to the next.

TYPICAL LIFESPAN
Not known

MAX. RECORDED AGE
11 years 10 months

The Redwing is primarily a winter visitor to Britain and Ireland, although we do have a tiny breeding population in northern Scotland. The arrival of these small thrushes signals the end of autumn, when their flight calls can be heard on late September evenings.

WINTER VISITORS
Our wintering Redwings come from two distinct breeding areas and belong to two races, which may be separated by those with a keen eye. The BTO's *Migration Atlas* shows that those wintering in England and across much of Scotland are predominantly from breeding populations in Finland and Russia. Those wintering in Ireland and parts of Scotland come from Iceland and are larger in size with duskier plumage. Redwings are nomadic in their winter movements, moving over large distances in response to food availability and weather conditions. This means that a bird feeding in a Surrey garden one winter, could be wintering in Greece or Italy the following year. This also explains why the BTO Garden BirdWatch reporting rate for this species varies so much from one year to the next. Ringing studies show that Redwings arrive in large numbers from October and then move into gardens. The Garden BirdWatch results also show this pattern, with garden reporting rates increasing from mid-October and peaking in late December or early January. In mild weather they will feed on the ground in open fields, taking earthworms, or in hedgerows taking fruits and berries. During cold weather they can be found in woodland, turning over leaves in a similar manner to that adopted by Blackbirds.

IT'S A FACT...
Redwings are nocturnal migrants and can be heard calling on still evenings after dark, as the migrating birds seek to remain in contact with one another.

IT'S A FACT...
Redwings have bred in Britain on a number of occasions, most notably during a period when the Finnish and Swedish populations peaked and the breeding range showed an expansion to the south.

IDENTIFICATION

The Redwing is much smaller than a Fieldfare. It has a bold white stripe above the eye (the only one of our thrush species to show this feature) and a less obvious one below the cheek. The stripe above the eye is known as the supercilium and is off-white in colour, extending from the bill, over the eye, and onto the nape. The two stripes, together with the rusty red colour under the wing and dark body tones, make separation from the slightly larger Song Thrush quite easy. Be aware of the rusty-buff colour that appears beneath the wing of the Song Thrush. The typical call is a thin 'tseep', uttered both at rest and in flight.

Earthworms are especially important in the breeding season.

COLD KILLS

Redwings are extremely vulnerable to the effects of very cold weather and, unable to find food, they can suffer from very high levels of mortality. During the very bitter winter of 1962/63 mortality levels were such that only the common Starling was more often reported found dead. This susceptibility to the effects of snow or hard frost underpins their nomadic nature, the birds forced further and further west as the weather deteriorates. Even though most of the Redwings arriving here in winter will have bred in areas undisturbed by Man, they remain extremely flighty and nervous birds, only overcoming their customary wariness if they are desperate for food. At such times they may venture into smaller gardens to take kitchen scraps, fruit and grated cheese.

A FLIGHT TOO FAR

In winter, during periods of cold weather, our Redwings may move further west into Ireland or southwest into Spain and Portugal. However, they don't always reach their intended destination. Redwing CX02630, ringed in a garden near Nuneaton, was found dead 3,900km away and three days later on a ship in the mid-Atlantic and heading towards Britain. This bird must have left the West Midlands in fairly good condition but missed Ireland by going just too far south.

While they last, Hawthorn berries are important food items.

THE MOVE INTO GARDENS

It is during the coldest weather that they venture into gardens, to feed on berries, and a flock of Redwings can strip a holly or Cotoneaster of its berries in just a few hours. Although Redwings are not regular visitors to feeding stations, they can be persuaded to take windfall apples and other fresh or dried fruit. In most years, fewer than 10% of Garden BirdWatchers report Redwings during any one week, but in January 1997, nearly 40% of observers reported them, corresponding with a period of very cold weather. During the winter, hundreds of Redwings may gather together to roost in tall conifers or thick shrubberies. The birds call as they arrive (don't confuse them with the noisy roosts of Starlings) but then settle quickly. At dawn, the birds depart in small groups, leaving discretely in search of feeding opportunities.

Berries on ornamental shrubs are increasingly important as winter progresses.

STATUS

Unfavourable conservation status within Europe, hence Amber listing.

SEASONAL TRENDS

BTO GARDEN BIRDWATCH REPORTING RATE (%) THROUGHOUT THE MONTHS OF THE YEAR

BTO BIRDFACTS

FAMILY	Turdidae
LENGTH	23cm
WINGSPAN	34cm
WEIGHT	83g
HABITAT	Woodland, gardens and farmland.
FOOD	Invertebrates, especially earthworms and snails. Fruit is taken in autumn and winter.
NEST	A neat structure of twigs, grass and some moss, lined with mud, dung and rotten wood, positioned in tree or shrub, often against the trunk. May also be positioned on a ledge or on the ground.
EGGS	3–5 eggs, pale bright blue, lightly spotted or speckled dark purple-brown.
EGG SIZE	27.4 × 20.8mm
INCUBATION	14–15 days. 2–4 broods.
BREEDING SEASON	March to September
POPULATION SIZE	1 million breeding pairs.
TYPICAL LIFESPAN	3 years
MAX. RECORDED AGE	10 years 8 months

SONG THRUSH
Turdus philomelos

The lawns and shrubberies of a typical garden are ideal for the Song Thrush, providing both feeding opportunities and thick cover for nesting. With Song Thrush populations known to be in decline in farmland, this makes gardens all the more important for the species.

GOING FOR A SONG

The Song Thrush is a popular songster. Even though it may lack the range of improvisation shown by the Nightingale or the rich serenity of a Blackbird, the Song Thrush has a song that has won the hearts of many listeners. The song itself is bold, with a bell-like clarity, and is structured through a series of phrases, many of which are repeated three times in succession (a useful identification feature). The full repertoire may consist of 100 or more different phrases and the bird seems to draw upon these almost at random. There is also a sense that some of the phrases have been copied from other birds or human sounds, like telephones or alarms.

Singing adult.

SONG THRUSH OR MISTLE THRUSH?

For some observers, separating the Song Thrush from its larger cousin can prove tricky. The smaller Song Thrush usually shows warmer tones and the spots on the breast differ from those of the Mistle Thrush in being inverted 'arrowheads' and much more sparsely spread on the breast and flanks. In Mistle Thrush the spots on the breast are more rounded, coalescing to form dark patches at the fold of the wing. While the 'tic' call of a Song Thrush is rather like the sound made by a freewheeling bicycle, the characteristic call of a Mistle Thrush is a harsh rattle (like a football rattle).

adult

juvenile

Adult in flight.

IN DECLINE !

The decline of the Song Thrush, which began in the mid-1970s, has been monitored by the BTO through its Common Birds Census and, more recently, the Breeding Bird Survey. The rapidity of the decline (some 50% over 38 years) earned the Song Thrush its place on the 'Red List' as a species of high conservation concern. During the early decades of the 20th century, Song Thrushes were more abundant than Blackbirds but since the 1940s Blackbirds have been the commoner of the two. In rural areas there are now nine Blackbirds for every Song Thrush and in urban areas it is more like 12 Blackbirds for every Song Thrush.

SLUG PELLETS

It has been suggested that the use of slug pellets is one of the reasons for the decline in Song Thrushes and, while we don't have any direct evidence that this is the case, many wildlife gardeners avoid their use, not least because they are toxic to other animals, including household pets. We know, from work carried out at the BTO, that increased mortality of young Song Thrushes, particularly in the first months after they fledge, is likely to have caused the population decline. What is not clear is why these youngsters are failing to survive. It may be that the increased use of slug pellets and lawn treatments has reduced the availability of snails or earthworms at a time of year when other foods are difficult to find. Away from gardens, changes in farming practices, land drainage and woodland management have all been put forward as potential causes for the decline.

With several hungry mouths to feed, parent Song Thrushes have their work cut out for them.

NESTING

The nest of a Song Thrush is fairly easy to recognise. As with other thrushes, mud is incorporated into the nest but, unlike other thrushes, the Song Thrush does not bother with a grass lining. Instead the lovely blue eggs are laid direct onto the smooth mud lining.

IT'S A FACT...

Although Song Thrushes like to nest in thick cover, they also like to be able to see out from the nest, enabling them to slip away unnoticed if danger threatens.

STATUS

Long-term decline in population has triggered inclusion on Red list.

SMASH AND GRAB

Many people associate Song Thrushes with the habit of smashing open snail shells. Interestingly, snails are only an important food source for Song Thrushes at certain times of the year. During the early part of the breeding season (March-May), Song Thrushes feed mainly on earthworms, but by early June they may have switched their attentions to caterpillars. It is only really in late summer that snails become important, as other foods become hard to find. This is particularly true in hot, dry spells when earthworms retreat deep into the soil. Snails can also be important during cold winters, when frozen ground restricts access to earthworms. The Song Thrush has developed a rather clever technique for dealing with snails. The bird usually grabs hold of the snail by the lip of the shell, before carrying it to a favoured stone, which it uses as an anvil. The impact typically only removes part of the shell but this still allows the bird to extract the snail's body, which is then wiped on the ground before being eaten. Other thrushes do not normally feed on snails in this manner.

Vision is the important sense for feeding Song Thrushes.

before   after

SEASONAL TRENDS

BTO GARDEN BIRDWATCH REPORTING RATE (%) THROUGHOUT THE MONTHS OF THE YEAR

BTO BIRDFACTS

MISTLE THRUSH
Turdus viscivorus

FAMILY	Turdidae
LENGTH	27cm
WINGSPAN	45cm
WEIGHT	130g
HABITAT	Woodland edge, suburban parks, large gardens.
FOOD	A range of invertebrates, with berries of particular importance in autumn and winter.
NEST	A cup, lined with a layer of mud sandwiched between an inner lining of fine grasses and an outer one of roots, grass and sticks.
EGGS	4 eggs, glossy greenish- or pale-blue, speckled with red-brown markings.
EGG SIZE	31.2 × 22.3mm
INCUBATION	15-16 days. 2 or sometimes 3 broods.
BREEDING SEASON	March to late June
POPULATION SIZE	205,000 pairs.
TYPICAL LIFESPAN	3 years
MAX. RECORDED AGE	11 years 4 months

Around the turn of the 19th century Mistle Thrushes were rare in northern Britain and Ireland. Progressively, the population increased and by the 1850s most of Ireland had been colonised, with Mistle Thrushes now found over most of Britain and Ireland and occupying woodland edge habitats, farmland and large gardens.

EARLY BREEDING

As a breeding bird the Mistle Thrush is an inhabitant of woodland edge, open country with scattered tall trees and large suburban parks. This, structurally the most impressive of our thrushes, is often heard in full song right at the start of the year. Indifferent to wind and rain, the Mistle Thrush will proclaim ownership of its breeding territory from the top of the tallest tree, a behaviour that has led to the local name of 'storm cock'. Egg-laying can begin as early as the end of February in mild winters but, with each pair rearing up to three broods of chicks, nesting activity may then continue through until the end of June. Adult Mistle Thrushes are surprisingly noisy around the nest, as are the family parties seen later in the year. The harsh 'football rattle' call is characteristic and should allow instant recognition.

adults

juvenile

IT'S A FACT...

Mistle Thrushes are more wary than other thrushes and your most likely view is one from behind as the birds fly off, uttering their rattling alarm call. You may see the white spots on the tail feathers.

BERRY IMPORTANT

The 18th-century Swedish biologist Linnaeus gave this bird the Latin name of 'viscivorus' ('devourer of mistletoe') when he translated the Greek word used by Aristotle to describe this thrush. Aristotle had watched Mistle Thrushes feeding on the red fruits of the mistletoe *Viscum cruciatum* around the Mediterranean. Today in Britain, Mistle Thrushes still feed on Mistletoe *V. album* (a white-berried species) but it is a less important food than the red berries of holly. One interesting aspect of Mistle Thrush behaviour is the way in which some individuals vigorously defend a supply of berries, typically on large, isolated hollies (but also hawthorn, Cotoneaster and large clumps of mistletoe). In some instances, both birds from a pair will defend a bush, driving away other birds and maintaining the fruit supply for themselves. During the course of protecting the tree and its precious berries, the birds themselves may also feed on nearby (unguarded) bushes or on the ground, effectively ensuring that their personal supply lasts as long as possible. Birds that guard berries have been shown to produce bigger and earlier clutches than those that do not guard berries. Because Holly berries can remain on a bush for many months without any serious deterioration, the berry supply can last through the winter and well into spring. The berries of other shrubs may not last as long.

Mistletoe berries

Holly berries

155

THE NEST

The nest is a large cup, placed in the fork of a horizontal branch or up against the trunk. Like those made by other thrushes, it has an outer lining of loosely woven grasses, moss and roots, held together with mud, leaves and rotten wood. The nest is completed with a lining of fine grasses and, occasionally, pine needles. It has been suggested that pine needles may help reduce the unwanted attentions of nest parasites. The outer layer is untidier than that seen in the other thrushes, often containing plastic or paper.

IN DECLINE !

Our Mistle Thrush population has undergone a significant decline since the 1970s, something that is thought to be linked to a fall in survival rates. However, the underlying cause remains unknown and, with a similar decline noted elsewhere in Europe, there is some cause for concern.

FAMILY BREAK-UP

Roving parties of Mistle Thrushes, a feature of late summer and early autumn, are often conspicuous, dropping into larger suburban and rural gardens. These groups of five or six birds are family parties, which will disband with the arrival of winter, the adults to defend their winter fruit supplies and the young to disperse. Autumn also sees the arrival of small numbers of Mistle Thrushes from breeding populations elsewhere in Europe. However, it is not known whether these birds remain here to overwinter or are merely stopping over on their way further south.

adult

adult

Adult foraging in snow.

IT'S A FACT...

Most Mistle Thrushes require deciduous trees in which to nest and as a vantage point from which to issue their loud song. This can be heard from late December, often in windy weather, hence the local name 'storm-cock'.

STATUS

Amber listed because of long-term decline.

SEASONAL TRENDS

BTO *Garden BirdWatch* reporting rate (%) throughout the months of the year

BTO BIRDFACTS

FAMILY	Sylviidae
LENGTH	14cm
WINGSPAN	20cm
WEIGHT	16g
HABITAT	Hedgerows and scrubby habitats.
FOOD	Mainly insects, especially beetles, bugs and caterpillars.
NEST	A deep cup, constructed of grass, roots and other plant material, placed in low bush, nettles or tall grass.
EGGS	4-5 eggs, pale blue or green with fine olive blotches.
EGG SIZE	18.6 × 13.9mm
INCUBATION	12-13 days. 1 brood.
BREEDING SEASON	April to August
POPULATION SIZE	931,000 pairs.
TYPICAL LIFESPAN	2 years
MAX. RECORDED AGE	6 years 1 month

WHITETHROAT
Sylvia communis

Although really a bird of scrubby habitats, the Whitethroat is an occasional garden visitor, most often putting in an appearance during the periods of spring and autumn migration.

IDENTIFICATION

The adult male Whitethroat has a pale grey head and a bright white throat, with a dull brown back and reddish-brown wings. In addition, the chest is flushed pink. Females and immature birds share the rusty-brown wings and dull brown back, and have an off-white throat and brown head.

A BIRD OF LOW COVER

The Whitethroat is among the more obvious warblers, the male often performing his scratchy song during a jerky display flight or scolding the observer with a pugnacious 'tchack' call reminiscent of two pebbles being brought rapidly together. This is a bird of low cover, of bramble thickets, stands of nettles and of hedgerow bases, hence the local name of 'nettlecreeper'. The first of our summering Whitethroats begin to arrive from mid to late April, occupying favoured hedgerows and patches of scrubby vegetation. The summer season is short and so these migrants quickly get down to the business of breeding, the first eggs being laid in May, normally in a clutch of four or five. The nest itself is almost always built within about 30cm of the ground. The male will build several unlined nests made of plant stems, roots and cobwebs, from which the female will select her preferred choice, which she then lines with fine grasses and hair. Both sexes share the task of incubating the eggs, though the female does the greater share of the work, being the only one to incubate during the night.

IT'S A FACT...

Although they feed mainly on beetles and the caterpillars of moths and butterflies, Whitethroats make use of small fruits and berries during the autumn. On their wintering grounds in the Sahel, they take insects and the berries of the Salvadora bush.

UPS AND DOWNS

Whitethroats are long-distance migrants and typically leave Britain and Ireland during August, arriving in wintering areas to the south of the Sahara from late September. This region of Africa is known as the Sahel and, by the time the Whitethroats arrive, the rainy season will have just ended. Other migrant warblers and flycatchers continue to follow the rains south but the Whitethroats remain within the Sahel and feed on berries and insects found in acacia scrub. Berries can provide a rich source of nutrients, especially sugars, used in the process of laying down the fat needed for migration. Finding enough food to make the return journey north can prove difficult and in some years a large proportion of our breeding Whitethroats fails to make it back to Britain and Ireland. The most dramatic 'crash' happened in 1968/69 when a drought in the Sahel reduced the British breeding population by 75%. Since then, there has been some recovery in numbers but not to pre-crash levels. Of the 276,000 Whitethroats ringed in Britain and Ireland, just six have been found in the Sahel during our winter. This shows how much more ringing is needed to find exactly where, within this region, our birds spend the winter.

A female Whitethroat feeding her young.

juvenile

Bathing and drinking are important for Whitethroats throughout their stay in our region.

FUELLING MIGRATION

Successful migration is, in part, dependent upon having sufficient fuel reserves (in the form of fat). Whitethroats and other small migratory birds often build up their fat reserves very quickly, perhaps adding 10-13% of their body weight each day during the period of fattening that takes place immediately prior to migration. Some of the increase comes from eating more food, with birds making the most of autumn's bounty. At the same time, many species make better use of the food they ingest, either through changes in their metabolism or by selecting foods that are easy to digest. Many fruits are ideal for this purpose; low in fibre, often abundant and easy to digest, they also tend to be rich in sugars (which help in the deposition of fat). Even with this extra fuel on board, these small birds may not have sufficient reserves to complete their journey in one go. Instead they may have to make one or more stopovers, halting to refuel before the next leg of their journey. Whitethroats, for example, are known to stopover in Portugal, where they have been recorded feeding on the red berries of *Daphne gnidium*.

IT'S A FACT...

Whitethroats will often make repeated short flights just ahead of an observer if disturbed from a linear feature such as a hedgerow.

STATUS

Population has fluctuated in response to Sahel droughts. Currently stable, although at a much lower level than in the early 1970s. Green listed.

SEASONAL TRENDS

BTO *GARDEN BIRDWATCH* REPORTING RATE (%) THROUGHOUT THE MONTHS OF THE YEAR

BTO BIRDFACTS

FAMILY	Sylviidae
LENGTH	13cm
WINGSPAN	22cm
WEIGHT	21g
HABITAT	Woodland with a well-developed shrub layer for breeding, gardens in winter.
FOOD	Insects and other invertebrates in summer, berries in autumn and bird table fare in winter.
NEST	Finely constructed cup, with 'translucent sides' of grass and herb stalks, lined with finer grasses, rootlets and hair. Usually situated low to the ground in bramble or other shrubs.
EGGS	4-5 eggs, mainly white, buff or olive in colour, with darker markings. Very variable.
EGG SIZE	19.6 × 14.8mm
INCUBATION	13-14 days. 1-2 broods.
BREEDING SEASON	April to July
POPULATION SIZE	916,000 breeding pairs.
TYPICAL LIFESPAN	2 years.
MAX. RECORDED AGE	10 years 8 months

BLACKCAP
Sylvia atricapilla

This common and widespread warbler shows a pronounced winter peak in its use of gardens, something that might seem a little odd for a species that has traditionally been regarded as a summer visitor to Britain and Ireland, but read on...

WINTER WONDERS

We know, from studies involving bird ringing, that Blackcaps forming our summer breeding population arrive in Britain and Ireland from early April (females arrive a week or so later than the males), to breed in scrubby woodland habitats, and depart again from late August. Recoveries of ringed Blackcaps suggest that most of our breeding population winters in southern Iberia and northwest Africa yet small numbers of Blackcaps have been recorded wintering in Britain for many years. The size of this wintering population has increased significantly over recent decades. Amazingly, these wintering birds are not simply individuals from our breeding population that have chosen to remain here (although some may remain). Instead, most are from the central European breeding population - birds that have migrated here in the autumn. Wintering birds, arriving during September and October, feed on natural foods in woodland and scrub before moving into gardens from late December.

male

IT'S A FACT...

As with a number of other summer migrants, arrival dates have tended to become progressively earlier since the mid-1970s, a pattern that may reflect climate change.

IDENTIFICATION

The Blackcap is instantly recognisable by the presence of a cap. This cap is black in adult males (hence the name) but is rufous-brown in adult females and a dull rufous-brown in young birds of either sex. Young birds undergo a partial moult in the autumn when they replace their body feathers and acquire the adult cap colour. This means that individuals seen at garden feeding stations in winter can be sexed on the colour of their cap, with young males often showing a few brown feathers mixed in with the predominantly black cap.

female

SONG

Because the Blackcap prefers dense cover, it is a species that is more often heard than seen during the breeding season. To the untrained ear, its rich, melodic song can be difficult to distinguish from that of the Garden Warbler. Eric Simms, author of an excellent book on warblers, described the song of the male Blackcap as being 'a pure rich warble with clean, musical intervals' that was 'less rapid, less even and less uniform than the outpourings of the Garden Warbler'. Nevertheless, for a sure identification it is always worth seeing the songster.

Male singing.

EVOLUTION IN ACTION

Blackcaps from the British and Irish breeding populations migrate south to southwest in the autumn, a direction that takes them down through France and Spain and across into North Africa. The general heading of individuals from the western part of the central European breeding population is also southwest but there is a large amount of variation in this heading such that some individuals take a more westerly route and reach our shores. Historically, those that did arrive here were likely to have found conditions unsuitable. However, over recent decades conditions may have become increasingly favourable, with milder winters and more people providing suitable food at garden feeding stations, helping these winter visitors to survive. The birds that have wintered here have been shown to arrive back on their central European breeding grounds some two weeks before those individuals wintering around the Mediterranean Basin, something that might offer them a competitive advantage and allow them to raise more youngsters. This may go some way to explaining how the tendency to winter in Britain has spread so rapidly through this particular component of the central European breeding population. It is evolution in action.

NESTING

The Blackcap is not a bird that you would expect to find nesting in a garden since it prefers mature deciduous woodland with a well-developed shrub layer. However, nesting Blackcaps have been recorded in overgrown hedgerows and in scrub within urban parks and larger gardens. A tangle of Bramble and nettles at the edge of an overgrown shrubbery or small woodland makes an ideal spot for a nest.

Female at nest.

IT'S A FACT...

Many Garden BirdWatchers report regular use of garden feeding stations by overwintering Blackcaps and it may be that this provision of food is one of the factors that has enabled the Blackcap to winter so successfully in Britain.

male

SECTION

Although warbler species are primarily insectivorous in their diet, the Blackcap is somewhat more catholic in its choice of food, feeding in the winter on fat, bread, fruit and even meat. Detailed work carried out by bird ringers has shown that Blackcaps feeding on such foods are able to maintain, or even increase, their body weight. One aspect of Blackcap behaviour often reported by BTO Garden BirdWatchers is the aggressive nature of these birds at garden feeding stations. Blackcaps will regularly chase other species away from suitable food sources. Interestingly, the ability of this species to take a wide range of different foodstuffs might be one reason why it is one of the earliest summer visitors to arrive.

female

STATUS

Long-term increase in abundance, hence Green listing.

SEASONAL TRENDS

20 %
16
12
8
4
0
J F M A M J J A S O N D

BTO GARDEN BIRDWATCH REPORTING RATE (%) THROUGHOUT THE MONTHS OF THE YEAR

BTO BIRDFACTS

CHIFFCHAFF
Phylloscopus collybita

FAMILY	Sylviidae
LENGTH	10cm
WINGSPAN	18cm
WEIGHT	9g
HABITAT	Open woodland, scrubby habitats and young plantations.
FOOD	Mostly insects and other invertebrates. May take fruit occasionally and nectar (especially prior to migration).
NEST	Woven, domed nest with a wide but shallow side entrance. On or low to ground in bramble or other cover.
EGGS	4-7 eggs, glossy white, with small number of purple-brown spots or speckles.
EGG SIZE	15 × 12mm
INCUBATION	13-14 days. 1 or 2 broods.
BREEDING SEASON	April to August
POPULATION SIZE	Some 800,000 breeding pairs, with a small but increasing wintering population.
TYPICAL LIFESPAN	2 years
MAX. RECORDED AGE	7 years 7 months

The small and rather plain looking Chiffchaff is only an occasional garden visitor, most commonly reported during the autumn. This is the time of year when young birds explore the wider landscape before beginning their migration south, through southern Europe and into North Africa.

CHANGE IN WINTERING BEHAVIOUR
Over the last 30 years or so there has been a noticeable increase in the number of Chiffchaffs wintering in Britain. Unlike the Blackcap, which is also wintering here in increasing numbers, wintering Chiffchaffs rarely visit gardens. Instead, almost all of our wintering Chiffchaffs remain within the wider countryside, favouring damp sites in the mildest parts of the country, especially the southwest. Numbers at some of these sites, which are often sewage farms or reedbeds, can top three-figures in some years. Such damp sites are likely to provide the insects and other invertebrates that these birds need to get through the winter. Although studies of ringed birds show that many of these wintering Chiffchaffs are British breeders, at least some originate from elsewhere in Europe since they show characteristics of other races. Continued studies will be needed to see what effect, if any, climate change has on wintering Chiffchaff numbers.

Being small and agile means that the Chiffchaff can get at those insects that live on the outer branches of trees. They are fairly opportunistic, taking whatever small insects and spiders are available.

IT'S A FACT...
Chiffchaffs are most numerous in the southern half of Britain, south of a line from the Wirral to the Wash. They are scarce in Scotland where they are greatly outnumbered by Willow Warblers.

Wintering Chiffchaff.

MOVEMENTS
We know from the work of specially trained and licensed bird ringers that many of our breeding Chiffchaffs spend the winter months around the western end of the Mediterranean and south into North Africa. Some winter to the south of the Sahara in Senegal and The Gambia.

Chiffchaff being ringed.

WHAT'S IN A NAME?

The Chiffchaff is one of the first migrants to return in spring and its song can be heard from the end of March across a range of scrubby woodland habitats. The species is also one of the few songbirds with an onomatopoeic name, in this case taken from its characteristic 'chiff-chaff' song. The scientific name of 'collybita' is also derived from the song; coming from the Greek word 'kollubistes', which means 'money-changer', the reference here is to the sound of coins being counted.

Chiffchaff

LOOK ALIKE

The Chiffchaff is similar in appearance to a number of other, closely related, leaf warblers, the most common of which is the Willow Warbler. Appearing smaller, and somewhat dumpy, the Chiffchaff has a somewhat drabber appearance than the Willow Warbler. It also has dark legs – Willow Warbler usually has pale legs and a brighter plumage with stronger overtones of green. The two species have very different songs, which helps with identification.

Willow Warbler

NESTING REQUIREMENTS

The Chiffchaff is a common breeding species in lowland woodland, favouring areas with a few taller trees for song posts and plenty of understorey vegetation in which to nest. The nest itself is typically built within a metre or so of the ground and often in bramble. The males, which return from their wintering areas a couple of weeks earlier than the females, quickly establish breeding territories and proclaim ownership through song. Increasing amounts of time are then spent in the understorey, searching for a prospective mate. Once a female is found, the male sets about trying to impress her of his suitability. He does this by floating down towards her with a slow butterfly-like flight, his legs dangling and his tail fanned and slightly raised. Shortly after this the pair bond is established, the male ceases his advertising song and the pair get down to the business of breeding. It is the female that selects the nest site, favouring thick cover, and who builds the nest. In fact, the male seems to lose interest at this stage and does very little to help; it is the female who incubates the eggs, broods the young and provides virtually all of the food needed by the growing chicks. It is interesting, though, that the male remains nearby throughout the breeding period, even if he actually contributes very little in the way of help.

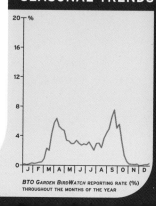

adult

Chiffchaff nest and young.

STATUS

Green listed as population has recovered strongly from a trough at the end of the 1970s.

SEASONAL TRENDS

BTO *GARDEN BIRDWATCH* REPORTING RATE (%) THROUGHOUT THE MONTHS OF THE YEAR

BTO BIRDFACTS

GOLDCREST
Regulus regulus

FAMILY	Sylviidae
LENGTH	9cm
WINGSPAN	14cm
WEIGHT	6g
HABITAT	Coniferous and mixed woodland.
FOOD	Spiders and insects (especially springtails, caterpillars and aphids).
NEST	An almost spherical cup of moss, lichens, feathers and cobwebs, usually suspended from twigs near the end of a conifer branch.
EGGS	6–8 eggs, white or pale buff, with fine grey-brown or purple speckles.
EGG SIZE	13.6 × 10.7mm
INCUBATION	16–19 days. 2 broods.
BREEDING SEASON	April to August
POPULATION SIZE	773,000 breeding territories.
TYPICAL LIFESPAN	2 years
MAX. RECORDED AGE	4 years 10 months

Goldcrests are birds of coniferous woodland, where their thin bills, agility and small size enable them to feed on the insects found at the very ends of branches, places where larger birds cannot forage. Some pairs breed in larger gardens where mature conifers are present.

OVERLOOKED

Although often overlooked, the Goldcrest is a widely distributed species that should be familiar to most observers. During the breeding season Goldcrests are most closely associated with coniferous woodland but at other times of the year they range more widely and often feature in gardens, where they can occasionally be seen feeding at bird tables on soft fat and bread. At this time of the year they associate with roving flocks of tits and other woodland birds like Treecreeper, seeking out suitable feeding opportunities. The Garden BirdWatch reporting rate is quite low (at about 3%) for much of the year but does peak at 9% in both spring and autumn (when migrants may be passing through the country). One other reason for the low reporting rate is that this species can be very difficult to spot. Its song and call note are very high pitched and often go unnoticed, especially when you get older and lose the high frequency component of your hearing.

male

male

IT'S A FACT...

As new conifer plantations mature they become increasingly suitable for Goldcrests, enabling the species to expand its population within new areas.

THE ROLE OF THE GOLD CREST

Goldcrests look busy as they forage on the outer branches of conifers, and often flick their wings as they move. The central crown stripe is orange-yellow in the male and lemon-yellow in the female. In both sexes this is flanked by lines of black. The male uses these markings to good effect during courtship, puffing out his feathers to display the orange crown to his mate. The males can be rather excitable, especially early in the breeding season when they are establishing their breeding territories. Two competing males will display at one another, again using the crest and presenting this towards their opponent by adopting a head down posture. At the same time the wings are lowered and each bird periodically adopts a strange bobbing movement as the display continues. Such display is usually sufficient to resolve the contest.

Male displaying.

CAUGHT OUT IN THE COLD

Goldcrests are known to be vulnerable to periods of severe winter weather. Their small size means that they chill more quickly than larger birds and they cannot store as much fat. To overcome this problem, they regularly roost together to conserve energy. They also use energy-saving behaviours during the breeding season. The nest is well insulated with a layer of feathers and the chicks minimise heat loss by burrowing down into the lower part of the nest once they have been fed. The effects of cold winters on our breeding population can be seen from BTO data gathered through the Common Birds Census, most notably following the 1962/63 winter. Goldcrests may be short-lived birds, with a high turnover of individuals, but they are able to rear a good number of young each year, if conditions are favourable, and this enables their populations to recover surprisingly quickly from the effects of cold winters. A recent run of mild winters has certainly helped this species.

COPING WITH WINTER

What makes the Goldcrest remarkable is that while some are migrants, undertaking long sea crossings to avoid the cold winters of northern regions, others remain in areas where darkness can last for 18 hours and temperatures fall to -25°C. Under such conditions, Goldcrests, huddling together for warmth, may still burn off 20% of their body weight overnight.

IT'S A FACT...

The Goldcrest is one of our smallest breeding birds; weighing in at just 6g it is half the weight of a Blue Tit.

Juvenile Goldcrests lack the crown markings and colours of adult birds.

SURPRISINGLY SMALL

Despite its small size and high energetic requirements, this remarkable little bird is amazingly productive. Each clutch of eggs (usually six to eight in number) adds up to a great deal of investment, often exceeding the female's own body weight. Some clutches contain 12 eggs, which is the equivalent of one and a half times the female's body weight. A large number of eggs usually follows through to a large brood of chicks, all of which require food. Despite having such a large brood, the female will usually start a second clutch in another nest while her mate continues to feed the first brood. It is this Herculean effort that underlies the Goldcrest's ability to bounce back from the effects of severe winter weather.

BELOW: **female;**
BOTTOM: **female at nest.**

STATUS

Amber listed because of longer term decline, though population within Britain currently increasing following run of mild winters.

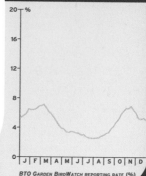

SEASONAL TRENDS

BTO Garden BirdWatch reporting rate (%) throughout the months of the year

SPOTTED FLYCATCHER
Muscicapa striata

Spotted Flycatchers are birds of woodland edge but are also found nesting in rural and suburban gardens. Garden nests are usually built against a wall, on a ledge or behind climbing plants, hence the local name of 'wall bird' used in some parts of England.

FAMILY
Muscicapidae

LENGTH
14cm

WINGSPAN
24cm

WEIGHT
17g

HABITAT
Open woodland and rural gardens.

FOOD
Hunts from a perch, taking flying insects.

NEST
A loosely built cup of dry grass, fine twigs and rootlets, lined with hair and feathers and placed on a natural or artificial ledge.

EGGS
4-5 eggs, pale-blue or off-white in colour and mottled red-brown or purple-grey.

EGG SIZE
18.5 × 14.1mm

INCUBATION
13-15 days. 2 broods.

BREEDING SEASON
May to August

POPULATION SIZE
Up to 59,000 breeding pairs.

TYPICAL LIFESPAN
2 years

MAX. RECORDED AGE
7 years 10 months

HERE TO NEST

The Spotted Flycatcher is one of the last summer migrants to reach us, typically arriving from the middle of May. Despite this late arrival, they may still manage to rear two broods of young before departing in late August. Spotted Flycatchers may be able to do this because they lay fewer eggs in their second clutch and because some of the young from the first clutch help to feed their younger siblings. The nest itself is almost invariably placed in excess of a metre above the ground, either in a shallow depression or against a wall. It may be built on top of the old nest of a thrush or Swallow and tends to be reasonably well-hidden. The Spotted Flycatcher is rather unobtrusive, drab in colour, with a soft squeaky song. However, they are often easy to spot because of their tendency to perch on an exposed post or branch, from which they make a darting flight to grab passing insects. Large flying insects are the preferred prey, particularly large flies, but the adults will take smaller insects from leaves and branches if the weather is cool and flying insects are unavailable. Because smaller insects tend to be eaten directly rather than being fed to the chicks, many Spotted Flycatcher nests will fail if the weather remains poor for a week or more. In summers with better weather, the birds are able to breed earlier and raise more chicks. When the female is forming her eggs she will often start to feed on woodlice and snails, both of which are rich in the calcium needed to form eggshells. The male will often support her at this stage by bringing extra food.

adult

ON THE MOVE

Our Spotted Flycatchers migrate south through western France and Iberia to winter in Africa. Ringing research shows that some of these winter in western Africa, south of the Sahara, but others may move farther south across the Equator to winter in South Africa. Difficulties encountered when migrating through the very dry Sahel region of Africa may have contributed to the 85% decline in Spotted Flycatcher numbers breeding in Britain since 1974. Recent BTO research suggests that other factors might also be involved, most notably a decline in the survival rates of young Spotted Flycatchers during their first year of life, maybe even during the first few weeks after leaving their nests.

Adult at nest.

IT'S A FACT...

If you are lucky enough to have Spotted Flycatchers nesting in your garden, set up some temporary perches nearby so that the birds can stop and check for predators before returning to the nest.

SOMBRE AND UNDERSTATED

The adult Spotted Flycatcher is a rather drab bird; dun-grey above and dull white below, the plumage lacks any obvious features. While it may not be much to look at, this demure little bird makes up for it by having a delightful character. Often confiding, Spotted Flycatchers will utilise a prominent perch from which they make short darting flights to snatch flying insects like flies, wasps and even bees. Remarkably, they are not deterred by stinging insects but handle them with exquisite poise. The sting is removed by rubbing the insect against the perch, a process that increases the amount of time needed to process the meal. If other prey is abundant, then the bird will avoid catching those stinging insects that need to be treated in this way.

adult

IT'S A FACT...

Young Spotted Flycatchers live up to their name, being much more heavily spotted than their parents.

juvenile

A SOFT SONG

You could almost be forgiven for thinking that the Spotted Flycatcher does not sing. Its squeaky, and largely unmusical, song is quite soft and not that far carrying. The song has two components, the squeaky components have a territorial function, while the sweeter phrases are aimed at potential mates.

CAPACIOUS GAPE

Spotted Flycatchers have a broad bill, which opens wide to help them catch flies. In addition, the bird has a series of modified feathers at the side of the bill, known as rectal bristles. The presence of these 'bristles' helps reduce the chances that its invertebrate prey might escape.

juvenile

STATUS

Declined by 85% since 1970s hence Red listing.

SEASONAL TRENDS

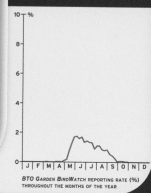

J F M A M J J A S O N D

BTO GARDEN BIRDWATCH REPORTING RATE (%) THROUGHOUT THE MONTHS OF THE YEAR

BTO BIRDFACTS

FAMILY	Paridae
LENGTH	12cm
WINGSPAN	18cm
WEIGHT	11g
HABITAT	Woodland, gardens.
FOOD	Insects and spiders, fruits and seeds in winter.
NEST	Pad of moss, mixed with dead grass, with a cup lined with hair, wool, fine grass and some feathers. Nest is placed in a suitable cavity or nestbox.
EGGS	8-10 eggs, white with fine red-brown spots usually concentrated at the broad end.
EGG SIZE	15.6 × 12mm
INCUBATION	13-15 days. 1, rarely 2 broods.
BREEDING SEASON	March to July
POPULATION SIZE	3.3 million breeding pairs.
TYPICAL LIFESPAN	3 years
MAX. RECORDED AGE	9 years 9 months

BLUE TIT
Cyanistes caeruleus

The Blue Tit's confiding and inquisitive nature make it one of the commonest visitors to garden feeding stations. In reality, the Blue Tit is primarily a bird of deciduous woodland, favouring oak woods because of the wealth of insects and spiders they support (ideal food for growing chicks) and the presence of suitable cavities in which to nest.

SUMMER STRUGGLERS

Despite the fact that they are so commonly reported from gardens, Blue Tits still find most gardens something of a difficult environment in which to raise a family. The average number of eggs laid in gardens (8.8 eggs) is fewer than that in mixed woodland (9.7) and mature oak woods (10.9). This is because gardens do not support anywhere near the number of insects (especially caterpillars) and spiders needed. Producing a clutch of eggs is very demanding and it has been calculated that a female needs 40% more food than normal during egg-laying. Adult Blue Tits can also have difficulty in finding enough food for their young. If you observe a Blue Tit moving through your rose bushes, picking off aphids, then the chances are that it has a brood of starving chicks to feed. The end result is that many young Blue Tits die in nestboxes, having starved to death. This lack of natural food appears to be compounded by the fact that the adults ignore much of the food provided on bird tables and in hanging feeders at this time of year. This is unsurprising, for while sunflower hearts and peanut fragments give the adults the energy they need at this busy time of year, such foodstuffs have a very different make-up from the energy-rich caterpillars fed to young chicks. Having said this, food on bird tables helps the adults to survive and may be fed to older chicks later in the breeding season. It also appears to be important once the chicks leave the nest.

adult

juvenile

Adult in flight, showing its rounded wings.

IT'S A FACT...

It seems that garden Blue Tits survive rather better than their country cousins, which means that the lack of production from garden nestboxes is compensated for by better survival of adult birds.

WINTER WONDERS

While gardens may be a poorer quality habitat than woodland when it comes to breeding, they can be very high quality habitats in winter, once food supplies in other habitats have run low. Garden feeding stations can attract as many as 200 individual Blue Tits in a single day, with a whole series of small flocks moving through and following a regular 'beat' that takes them from feeding station to feeding station. Although such flocks are mobile, they typically only range over a small area but during the course of a winter more than 1,000 individuals may visit a single bird table. We know this staggering piece of information because of the work of BTO ringers who regularly catch and ring Blue Tits in gardens during the winter.

With a covering of snow, food provisions in the garden assume an even greater importance for Blue Tits than at other times.

THE NEED FOR CALCIUM

The shell of a Blue Tit egg weighs about 0.1g and is mainly composed of calcium carbonate (about 50% of which by weight is calcium). A female therefore needs to find something like 0.5g of calcium to complete an average clutch of eggs. In some bird species, like the domestic hen, calcium is laid down in advance of the breeding season so that it is available when the eggs are to be produced. However, given that the total amount of calcium in a Blue Tit skeleton is only about 0.6g, there is no scope for a female Blue Tit to store calcium in this way. Every day, the female Blue Tit has to find 0.1g of calcium so that she can lay an egg the next morning. It seems that she manages to do this by deliberately feeding on fragments of old snail shells and, possibly, on bits of eggshell.

OPPORTUNISM

Blue Tits are seemingly very bright and resourceful birds and this has led to reports of individuals pecking at linseed-oil based putty or pecking through foil milk bottle tops to reach the cream. This adaptability is an important component of Blue Tit behaviour and may well have contributed to its success in exploiting the human environment and the feeding opportunities that we provide.

WINTER VISITORS

Scandinavian Blue Tits undertake regular long distance movements, occasionally resulting in eruptions of these more brightly coloured continental birds south and west into England. One such eruption occurred in 1957, when many continental Blue Tits were noted in south and eastern England. Such eruptions are rare today, perhaps because winters are now less severe than they were or because more people in Scandinavia now provide supplementary food throughout the winter months.

Adult in spring.

LARGE BROODS

British and Irish Blue Tits only raise one brood a year and adopt an 'all your eggs in one basket' approach, producing some of the largest clutches of any song bird. Most contain between seven and 13 eggs, though clutches of 15 or even 16 are not unknown.

adult

IT'S A FACT...

Autumn is the time when the fewest Blue Tits visit gardens but, come November, visiting numbers increase.

With 10 young mouths to feed, these parent Blue Tits have really got their work cut out.

Juvenile begging for food.

STATUS

Long-term general increase in abundance. Green listed.

SEASONAL TRENDS

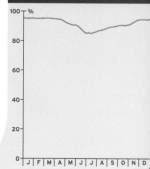

BTO GARDEN BIRDWATCH REPORTING RATE (%) THROUGHOUT THE MONTHS OF THE YEAR

BTO BIRDFACTS

FAMILY
Paridae

LENGTH
14cm

WINGSPAN
24cm

WEIGHT
18g

HABITAT
Woodland, suburban and rural gardens.

FOOD
Invertebrates, especially beetles and caterpillars. In winter Great Tits take a wider range of seeds and fruits.

NEST
Mat of moss and dry grass, cup lined with wool, hair and feathers. Placed in a cavity or nestbox.

EGGS
7–9 eggs, white eggs, variously speckled red-brown often with concentration of markings at broad end.

EGG SIZE
17.5 × 13.5mm

INCUBATION
13–15 days. 1 brood.

BREEDING SEASON
March to July

POPULATION SIZE
2 million breeding pairs.

TYPICAL LIFESPAN
3 years

MAX. RECORDED AGE
13 years 11 months

GREAT TIT

Parus major

Arguably the most handsome of our tit species, the Great Tit is also the largest. This, coupled with its habit of feeding on the ground, makes the Great Tit a very adaptable species and one that can be found across a wide range of woodland and semi-open habitats.

STATUS SIGNALS

The black stripe running down the breast plays an important role in Great Tit behaviour, acting as a badge denoting the status of the individual. In females, the stripe peters out about halfway down the belly, while in males it continues down between the legs and is both broader and bolder than in the female. There is less variation between members of the same sex but these subtle differences appear to be enough to show the status of the bird, those with bolder stripes being the more dominant. This 'badge' is shown-off most effectively when birds adopt a head-up display that serves to exaggerate the size of the stripe. It makes sense for birds to have some way of displaying their status to other individuals, since it helps to avoid unnecessary fights. The social system in Great Tits gives rise to a dominance hierarchy and it is the younger (and less dominant) individuals that are forced to feed on less profitable or higher risk feeders. Research has shown that dominant Great Tits choose feeders closest to cover, while younger birds feed at more exposed feeders, where the risk from Sparrowhawks is greater. Many other garden birds, e.g. House Sparrow, also have dominance hierarchies, often accompanied by some 'badge' used to denote status.

FEEDING TECHNIQUES

Great Tits use their feet to hold large food items, like Beech mast and acorns, while hammering these seeds open with their bill. A similar approach is used to kill caterpillars, a very important prey during the late spring and summer months. Dr Andy Gosler, who has studied Great Tits for many years, once calculated that the effort of feeding chicks was equivalent to a human parent bringing home over 100kg of shopping every day for three weeks! As summer moves into autumn, Great Tits begin to take more tree seeds. During the winter months, as seed supplies begin to run low, woodland populations of Great Tits become more mobile and often visit gardens to feed on the food we provide. One finding from recent research is that, as adult Great Tits switch from feeding on seeds in winter to feeding on caterpillars in spring, their bills become longer and less deep. Such changes are very small but appear to be enough to make the bill better suited to a diet of invertebrates – a short, robust bill being better for dealing with nuts and seeds.

Adult male Great Tit seen in profile (LEFT) and head on (RIGHT).

INCUBATION

The female Great Tit roosts in the nesting cavity once the first egg has been laid, but, even though she is in contact with them, she is not generating enough heat to start the incubation process. This only begins once all the eggs have been laid. The heat needed to allow the eggs to develop comes from the female through her brood patch. This is an area of bare skin that develops just prior to incubation, rich in blood vessels and hot to the touch. The female is able to maintain the surface temperature of the eggs at about 35.4°C by altering the amount of time she spends incubating them. Typically, this involves 30 minutes of incubation followed by a 10 minute break.

IT'S A FACT...

Great Tits also show a willingness to utilise nestboxes, maybe because natural cavities with a large enough entrance hole are hard to find.

Adult Great Tit broods its clutch of young birds.

IT'S A FACT...

Great Tits like bigger and deeper boxes than Blue Tits and need the entrance hole to be at least 28mm in diameter.

FINDING YOUR FEET

Once they leave the nest, young Great Tits are reliant on their parents for another couple of weeks. During this period the family party will travel some distance to find food and often visit bird tables, where the food we provide may be very important, since starvation is known to be a major cause of death.

Adult male singing.

juvenile

SING A SIMPLE SONG

The Great Tit's bold, loud, ringing song is familiar to most, if not all, garden birdwatchers. The most familiar call is the 'teacher teacher' call, often heard from late winter. Male Great Tits often develop a number of variations on the basic song but these have to be acquired, meaning that older birds are more diverse in their repertoire than young ones - one older male was found to use some 40 different songs. The amount of time spent singing declines once nesting starts (usually in April) and the female begins incubation.

PUTTING ON WEIGHT

When they emerge from the eggs, each chick weighs just over a gramme and, over the next two weeks or so, will dramatically increase in weight to some 15 times that at hatching. How well the developing chick does is dependent upon food availability, weather conditions and competition from other species. Great Tits raised in gardens are smaller than those in woodland, reflecting the reduced availability of caterpillars.

STATUS

Steady increase since 1960s, perhaps linked to increasing amounts of garden feeding. Green listed.

SEASONAL TRENDS

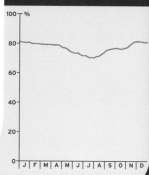

BTO GARDEN BIRDWATCH REPORTING RATE (%) THROUGHOUT THE MONTHS OF THE YEAR

BTO BIRDFACTS

MARSH TIT & WILLOW TIT
Poecile palustris & Poecile montanus

As well as looking very similar, Marsh and Willow Tits have much else in common. Both are infrequent visitors to gardens, both are highly sedentary in nature and both are causing concern amongst conservationists. Set against these similarities are many differences. Contrary to their name, Marsh Tits are associated with drier habitats and are most commonly found in larger patches of broad-leaved woodland with a well-developed shrub layer, while Willow Tits prefer damp areas or conifers.

FAMILY
Paridae

LENGTH
12cm (both species)

WINGSPAN
19cm (both species)

WEIGHT
12g (both species)

HABITAT
Marsh Tits like drier habitats, such as broad-leaved woodland with a well-developed shrub layer. Willow Tits prefer damp areas or conifers.

FOOD
Both species feed on insects and spiders, with seeds taken in winter, Marsh Tit also taking some nuts and berries.

NEST
Both use a hole in tree or stump but only the Willow Tit excavates its own hole.

EGGS
7-9 eggs (MT), 6-8 eggs (WT), similar looking – white with a few sparse red-brown spots at one end.

EGG SIZE
16.2 × 12.4mm (MT); 15.5 × 12.2mm (WT)

INCUBATION
14-16 days. 1, sometimes 2 broods (MT); 14 days, 1 brood (WT).

BREEDING SEASON
March to July (MT); April to July (T)

POPULATION SIZE
53,000 breeding pairs (MT); 8,500 breeding pairs (WT)

TYPICAL LIFESPAN
Too few birds ringed to be certain for either species.

MAX. RECORDED AGE
10 years (MT); 9 years (WT)

IDENTIFICATION

The black cap of the Marsh Tit is glossier than that seen in the Willow Tit and it does not extend so far down the nape. The Willow Tit's bib is larger and less sharply defined than that of the Marsh Tit. A useful feature worth looking out for in the Willow Tit is the pale cream edges to its wing feathers, which gives the appearance of a pale wing panel in a bird perched with its wings closed.

Marsh Tit

Willow Tit

SPOT THE DIFFERENCE

The Willow Tit has the unique distinction of being the last regularly breeding British bird to be identified and named, something that is perhaps unsurprising given the great difficulty in distinguishing this species from the very similar Marsh Tit. The similarity of their plumage means that these two species are difficult to separate in the field unless you are fortunate enough to hear (and to recognise) their distinctive calls. That of the Willow Tit is a characteristic 'zee-zurzur-zur', very different from the 'pitchou' of the Marsh Tit. Many aspects of the biology of these two species remain undiscovered, including a complete picture of their distribution across Britain. Both are known to be absent from Ireland, the Isle of Man and much of Scotland, although the Willow Tit does breed further north into Scotland than its slightly larger relative, extending into Strathclyde, Dumfries and Galloway and into the Borders and Lothian.

Marsh Tit

Willow Tit

Willow Tit

Marsh Tit

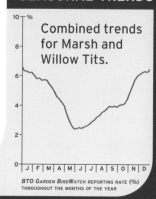

NESTING HABITS

Willow Tits excavate their own nest holes, typically in rotten stumps and within a metre or so of ground level. Because of this they are difficult to tempt into a nestbox, although this can be done by filling the box with wood chippings or a similar material (e.g. a polystyrene block), as demonstrated by Jimmy Maxwell in Scotland. The preferred trees are birch, willow and alder. It is worth leaving a rotten stump in the hope that a pair of Willow Tits will use it for a nest site. While the Willow Tit excavates its own nest hole, the Marsh Tit only ever goes as far as enlarging somebody else's. Being slightly larger and more aggressive, it is not unusual for a Marsh Tit to take over a hole excavated by a Willow Tit. Interestingly, Marsh Tits do not appear to take as readily to nestboxes as one might expect - though this could be due to competition from Blue and Great Tits.

IT'S A FACT...

BTO census work shows that there has been a rapid decline in the Marsh Tit population. There has also been an increase in the use of garden feeding stations, which suggests that natural food supplies are becoming scarce in winter.

FORAGING BEHAVIOUR

Marsh Tits hold very large territories for such small birds (some 2-3ha in size, which, by comparison, is about three times as large as that of a Great Tit) and both the male and female will remain there throughout the year. Younger, unpaired individuals will join up with mixed flocks of tit species in the winter and rove in search of food. During autumn and winter, Beech mast and other tree seeds become an important food source, something that these small birds can exploit because of their surprisingly strong bills. Like Coal Tits, Marsh Tits store food items, although only for a few days or even hours. Experiments have highlighted the amazing ability that these tits have to remember where they have hidden seeds. Willow Tits also form small foraging flocks in the autumn, though this behaviour is more prevalent in the northern part of the Willow Tit's range.

Marsh Tit

STATUS

Both species are Red listed because of decline in population size.

Marsh Tit

Willow Tit

SEASONAL TRENDS

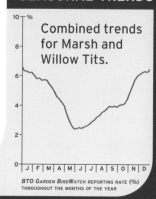

Combined trends for Marsh and Willow Tits.

BTO GARDEN BIRDWATCH REPORTING RATE (%) THROUGHOUT THE MONTHS OF THE YEAR

Marsh Tit

Willow Tit

COAL TIT
Periparus ater

BTO BIRDFACTS

FAMILY
Paridae

LENGTH
12cm

WINGSPAN
19cm

WEIGHT
9g

HABITAT
Coniferous and broad-leaved woodland, gardens.

FOOD
Insects and spiders, plus small seeds in winter.

NEST
A cup of moss, lined with hair and wool, in a cavity or nestbox.

EGGS
9-10 eggs, white, finely spotted or speckled red-brown.

EGG SIZE
14.9 × 11.6mm

INCUBATION
14-16 days.
1-2 broods.

BREEDING SEASON
March to July

POPULATION SIZE
604,000 breeding pairs.

TYPICAL LIFESPAN
2 years

MAX. RECORDED AGE
8 years 9 months

Typically less obvious than Blue Tit or Great Tit, the Coal Tit often flies in to a feeding station to grab a seed before disappearing off into a bush or shrub to feed. This behaviour reflects the fact that this is the smallest of our true tits, weighing in at just 8-10g, and is thus at the bottom of the 'pecking order'.

EVERGREEN PREFERENCE

Away from gardens, the Coal Tit tends to forage in the upper parts of large conifers, where it can be seen moving about on needles, thin branches and cones. The agility of this species is evident when foraging, with individual birds feeding upside down and hanging from the smallest branches. Coal Tits will also hover in much the same manner used by Goldcrests. The Coal Tit shows a number of adaptations for living and foraging within coniferous woodland. It has a fine bill, long toes and claws and its small size makes it the most agile of the tits. In Britain, the Coal Tit is more closely associated with conifers than other tits (with the exception of Crested Tit) but, interestingly, it makes greater use of broad-leaved woodland here than elsewhere in its European range. Those Coal Tits feeding within broad-leaved woodland will often feed on some of the larger branches and in this habitat they may spend up to 90% of the daylight hours feeding. The widespread establishment of conifer plantations has been of great benefit to the Coal Tit, with BTO surveys charting an increase in its population over the period that such plantations have matured. Some idea of the success of the Coal Tit in such habitats can be seen from the fact that densities of up to 50 pairs/km^2 have been recorded in Scots Pine plantations within the Norfolk and Suffolk Brecks. Coal Tits also make use of other habitats and are regular visitors to gardens, parks and cemeteries, especially if there are ornamental conifers present.

WHAT'S IN A NAME?

Until recently all of our 'true' tits (Blue, Coal, Great, Marsh, Willow and Crested Tit) were placed together in the genus *Parus*. This placement was based of structural characteristics. However, structural characteristics may be misleading because two species may evolve similar features, yet still be quite unrelated. Recent advances in our understanding of DNA have enabled us to reassess how species are related to one another and, from time to time, this results in a need to reclassify species. The net result of this, following work on our tit species, is that while Great Tit remains in the genus *Parus*, the other species move into new genera and so Coal Tit becomes *Periparus*.

Adult foraging in conifers.

IT'S A FACT...

The white nape patch is a useful identification feature, helping to separate this species from the two other black-capped tits, Marsh Tit and Willow Tit.

adult

adult

adult

THE IMPORTANCE OF SEED

Seeds are an important part of the diet during winter. Research has shown that the availability of both conifer seeds and Beech mast has a pronounced influence on the degree to which Coal Tits will use gardens. Gardens are more likely to be used in those years when the seeds of Sitka Spruce and Beech are in short supply. Elsewhere in Europe, Coal Tits are known to make long distance movements in search of food during years when tree seed may be in short supply. In some years, such movements may bring small numbers of continental Coal Tits to southeastern Britain, as exemplified by a number of recoveries of German and Dutch-ringed individuals in England.

PROVISIONING

Coal Tits are well known for their ability to store food and can be seen taking seeds away from a feeding station to cache for use at a later date. This behaviour, coupled with the fact that they can feed on the undersides of snow-covered conifer branches, enables them to survive severe winter weather more readily than either Blue Tits or Great Tits. Food caching usually occurs between June and December, with a peak in October, but there have been observations that indicate Coal Tits may also store food in April during the run-up to egg-laying – a period when additional food resources may be particularly important.

IT'S A FACT...

The provision of a nestbox with an entrance hole of no more than 25mm is the best option if you wish to attract a breeding pair and prevent them being evicted by other species.

Adult showing its extensive black 'bib' and relatively powerful feet and legs.

BREEDING BEHAVIOUR

The breeding season begins in March, with new breeders pairing up following the break-up of winter foraging flocks. In Britain, where the species is resident, established pairs will remain together from one year to the next, associating with each other in the winter flocks. Males and females visit potential nest holes together, although it is believed that it is the female that ultimately decides which site is to be used. Coal Tits prefer to use nest sites with a small, narrow entrance. No doubt, this helps them to avoid competition with their larger relatives. If they select a site to which Blue Tits or Great Tits can gain access then there is a danger that they will be evicted. The nest itself is built by the female alone and she is also responsible for incubating the eggs and brooding the chicks. Once the chicks are large enough to be left on their own, both sexes will provision them, bringing in small insects and spiders. The pair bond may be strengthened through courtship feeding, where the male provides food for the female, a behaviour seen in a range of bird species. Coal Tits often manage to rear a second brood of chicks later in the season, something that Blue Tits and Great Tits manage less often.

By the time the young are close to fledging, a parent bird's plumage looks decidedly moth-eaten, the result of wear and tear while foraging for food.

STATUS

Population rather stable since 1970s. Green listed.

SEASONAL TRENDS

BTO GARDEN BIRDWATCH REPORTING RATE (%) THROUGHOUT THE MONTHS OF THE YEAR

BTO BIRDFACTS

FAMILY
Aegithalidae

LENGTH
14cm

WINGSPAN
18cm

WEIGHT
9g

HABITAT
Deciduous woodland, scrub and larger gardens.

FOOD
Largely insectivorous, but will take fat (especially if smeared onto branches), small seeds, finely-grated cheese and bread crumbs from garden feeding stations.

NEST
An impressive dome, made from mosses and lichens, and lined with feathers.

EGGS
6-8 eggs, glossy white, either unmarked or with tiny red speckles.

EGG SIZE
14.2 × 10.8mm

INCUBATION
15-18 days. 1 brood.

BREEDING SEASON
March to June

POPULATION SIZE
261,000 breeding territories.

TYPICAL LIFESPAN
2 years

MAX. RECORDED AGE
8 years 8 months

LONG-TAILED TIT
Aegithalos caudatus

Despite its name, the Long-tailed Tit is only distantly related to the familiar Blue and Great Tits, with which it often associates. It does not nest in a cavity but instead constructs a delicate domed nest, spherical in shape and placed either low down in dense bushes or high up in the fork of a tree.

AN APPEALING BIRD
The appearance and behaviour of this species make it one of the most delightful of garden visitors, attracted to peanuts, fat and other fare at garden feeding stations. Garden BirdWatch figures show that usage of gardens peaks in late winter, especially February and March, when food may be scarce in other habitats. Being such small birds, Long-tailed Tits are susceptible to cold winter weather. This susceptibility explains why numbers can fluctuate so dramatically, with the breeding population sometimes reduced by up to 80% following a particularly severe winter. Fortunately, recent winters have been relatively mild, prompting an increase in numbers.

INCREASING USE OF GARDENS
A recent feature of Long-tailed Tit behaviour has been the increased use of gardens. Garden BirdWatchers have noticed these birds adapting their behaviour to begin feeding from peanut feeders. Historically, Long-tailed Tits have remained more strongly insectivorous than other tits, although there are also records of them feeding on fat, small seeds, bread and cheese crumbs. These are taken from hanging feeders, bird tables and from the ground. It appears that, once one individual within a group learns to exploit a new food source, the other members soon adopt the behaviour. This is another clear benefit of group living.

APPEARANCE
Long-tailed Tits have small black, white and pink bodies, with a long tail. Young birds are duller in appearance than their parents and lack the pink colouring completely. Because the young moult shortly after leaving the nest, they soon look indistinguishable from their parents. Birds from some other parts of Europe look distinctly different from those found in Britain and Ireland – for example, we are occasionally visited by individuals from northern Europe, distinguishable by their all white heads.

Long-tailed Tit Scandinavian race

Long-tailed Tit British race

Adult, fluffed up against the cold in winter.

IT'S A FACT...
Away from gardens, Long-tailed Tits prefer to forage in the canopy of taller trees and shrubs, searching for small insects and the eggs and larvae of moths.

AMAZING NESTS

Long-tailed Tits build elaborate domed nests, often placed in the middle of thorny bushes. The nest itself has an outer structure, which is then lined with up to 2,600 feathers, a process that may take 39 days to complete. The birds prefer feathers that are between 2 and 4cm in length and, judging from experiments that have been carried out on feather availability and use, there does not seem to be any shortage of these. One of the most amazing aspects of the nest building process has only been discovered within the last few years. It seems that Long-tailed Tits use the feather lining to regulate the temperature within the nest and are able to accurately gauge how many feathers they need to get the temperature they want. Researchers discovered this by adding feathers to nests that were already under construction and then recording how many more feathers were brought in by the birds themselves. The birds were able to adjust the number of feathers they collected to accommodate the additional ones that had been added by the researchers.

A Long-tailed Tit's nest is a miracle of construction and extremely well camouflaged in dense vegetation.

AN EXTENDED FAMILY

Nest building often starts early in March and may take several weeks to complete, hardly surprising given that the nest is so finely constructed. Adults whose own breeding attempts have failed may act as helpers at the nests of their relatives. The behaviour of this species has been very well studied and we now understand the complex interactions that take place between the various members of the small parties that may gather together in the winter. These often include the parents and young from several nests, together with adult birds that had failed to rear chicks during the breeding season. Many of the birds in these parties will be related and together they defend a winter feeding territory. They also roost communally, a trait that makes a very important contribution to being able to survive the cold winter nights. Long-tailed Tits do not roost in holes, but huddle close together, often in a line along a branch in the middle of a bush. Towards the end of the winter, unpaired females leave the group and join a neighbouring party. The remaining pairs start to build nests within their winter range, continuing to roost communally until the dome of the nest is complete.

IT'S A FACT...

Choosing a thorny site for nesting seems to offer some protection from the very high levels of nest predation that these birds suffer, often by Jays or Magpies.

This adult is investigating a thorny bramble patch for its suitability as a nest site.

Adult birds are flushed pink on their underparts and back.

STATUS

Although the population undergoes wide fluctuations in numbers it is currently Green listed.

SEASONAL TRENDS

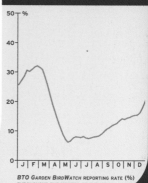
BTO GARDEN BIRDWATCH REPORTING RATE (%) THROUGHOUT THE MONTHS OF THE YEAR

BTO BIRDFACTS

NUTHATCH
Sitta europaea

Nuthatches make ready use of natural tree holes for nesting.

FAMILY	Sittidae
LENGTH	14cm
WINGSPAN	24cm
WEIGHT	24g
HABITAT	Woodland and larger gardens.
FOOD	Invertebrates, mainly taken from tree trunks, but also seeds (especially in autumn and winter).
NEST	Hole in tree or, sometimes, a nestbox.
EGGS	White in colour, sparsely speckled with red-brown.
EGG SIZE	19.3 × 14.8mm
INCUBATION	16–17 days. 1, sometimes 2 broods.
BREEDING SEASON	March to June
POPULATION SIZE	144,000 territories.
TYPICAL LIFESPAN	3 years
MAX. RECORDED AGE	11 years 9 months

This is a spectacular visitor to any bird table, not least in Scotland where the species was first recorded breeding in 1989. The Nuthatch does not breed in Ireland and, given its sedentary nature, seems unlikely so to do.

ON THE UP

Nuthatches are birds of broad-leaved and mixed woodland but will use large, well-timbered, gardens, especially when situated near to blocks of suitable woodland habitat. They are surprisingly anti-social birds, with pairs vigorously defending their territories throughout the year. Even newly-fledged young can be aggressive towards each other. By holding a territory throughout the year, Nuthatches become familiar with what food is available and where it has been stored – a very useful survival strategy. Nuthatch populations have increased rapidly since the mid-1970s and there is, as yet, no sign of a halt to this upward trend. Alongside this increase in abundance there has also be a northwards expansion of the breeding range within Britain. The Nuthatch has spread into southern Scotland and increased its populations in northern England. The reasons for the increase are unknown but BTO researchers have identified a tendency towards breeding earlier in the year, together with a large increase in brood size. Could such changes be the result of climate change?

BREEDING BEHAVIOUR

Breeding starts early, with males singing from December, and the pair quickly selects and defends a nest hole in a mature tree. The size of the entrance hole is not as crucial as it is for some other species because the Nuthatch uses mud to reduce the size of the aperture to a point where the female can just squeeze through (about 3cm). Nuthatches do not take readily to nestboxes but, where they do use them, they will often either cement the roof of the box shut or cement the box to the tree, in addition to reshaping the entrance hole.

IT'S A FACT...

The chisel-like beak is not strong enough to drill into wood in the way that a woodpecker does, but it does enable the Nuthatch to remove pieces of bark or access rotten wood.

An adult bird investigating a natural tree hole for food.

POSSESSION IS NINE-TENTHS OF THE LAW

As already noted, Nuthatches are overtly territorial birds and the monogamous pairs will usually remain on their territory throughout the year. Soon after the young have fledged they will be driven away from the territory in which they were born to occupy vacant territories elsewhere. Researchers have shown that older females tend to lay their eggs slightly earlier in the season than young birds, so it is their young that are the first to occupy these vacant territories. Interestingly, these early youngsters tend to retain the territories they have first secured even though you would think that later arriving youngsters would try to dislodge them. That they hang onto these territories suggests that, for Nuthatches, possession is nine-tenths of the law.

FOODS AND FEEDING

Insects and spiders are usually gleaned from the trunks of trees but in autumn and winter various tree seeds (especially Beech mast, acorns and hazel nuts) are favoured. These are often wedged into cracks and crevices where they are then smashed open by the strong chisel-like bill, a behaviour that has given rise to local names like 'nuthacker', 'jobbin' and 'nut jobber' ('job' is an old English word meaning to stab with a sharp instrument). The Garden BirdWatch reporting rate for this species is influenced by the size of the Beech mast crop. In years with a good crop the reporting rate tends to drop. In autumn 2002, and again in autumn 2006, the Beech mast crop was widely in short supply and this resulted in a dramatic increase in the number of gardens in which this species was reported, as foraging adults and dispersing young roamed more widely. Nuthatches will feed on peanut feeders, often dominating them to the exclusion of most other species. Nuthatches are great hoarders of food and can sometimes be seen making repeated trips to a feeder, emptying it of its contents that are then cached locally. One enterprising Nuthatch hid its cache under a pillow in the guest bedroom of one of the BTO's Garden BirdWatch Team! The small window to this room was always left open.

IDENTIFICATION

Steely blue grey above and pinky buff below, this species has a striking black eye-stripe and chisel-like bill. It can often be seen climbing down tree trunks headfirst, a behaviour not used regularly by any other British bird. Males have a stronger red-brown (virtually brick-red) colour to their flanks than females.

Adult bird climbing down a tree trunk.

A Nuthatch visiting a garden bird feeder.

STATUS

Increasing, so Green listed.

SEASONAL TRENDS

20 %
16
12
8
4
0
J F M A M J J A S O N D

BTO *Garden BirdWatch* reporting rate (%) throughout the months of the year

BTO BIRDFACTS

TREECREEPER
Certhia familiaris

FAMILY	Certhiidae
LENGTH	12cm
WINGSPAN	19cm
WEIGHT	10g
HABITAT	Woodland, broad-leaved or coniferous in nature.
FOOD	Insects and spiders taken from crevices in bark. Conifer seeds also taken in winter.
NEST	A cup of grass, rootlets and moss, wedged in behind a loose flap of bark.
EGGS	5-6 eggs, white with very fine pink-brown speckles.
EGG SIZE	15.6 × 12.2mm
INCUBATION	13-17 days. 2 broods.
BREEDING SEASON	April to July
POPULATION SIZE	204,000 breeding territories.
TYPICAL LIFESPAN	2 years
MAX. RECORDED AGE	8 years

The unobtrusive Treecreeper is easily overlooked; its dappled plumage is brown and black, and it only occasionally comes to bird tables (though it will sometimes come to peanut feeders and fat smeared on trunks).

HARD TIMES

Like other small birds, the Treecreeper can face difficulties during severe winter weather. In particular, glazing frosts or freezing rain can prevent Treecreepers from reaching bark-dwelling insects and their larvae. Treecreepers will seek shelter by roosting in a cavity or behind a flap of bark. Roosting pockets may be used and are worth attaching to the trunks of larger trees – place them some 3m above the ground.

TREE MOUSE

The local West Country name for the Treecreeper is 'Tree mouse' and it is easy to see why. This bird climbs up tree trunks in a 'mouse-like' fashion supported by its stiff tail. The Garden BirdWatch reporting rate, which averages just under 20%, may under-emphasise their use of gardens. Results suggest that more gardens contain Treecreepers between November and April, although they may be more conspicuous at this time because many trees are devoid of their leaves and the birds themselves are at their most vocal. Even so, the song and calls are high-pitched and are sometimes missed.

DISTRIBUTION

Within Britain, the Treecreeper is most at home in broad-leaved woodland but this is not the case elsewhere in Europe, where it is a bird of upland conifer forests. Lowland broad-leaved woodland in Europe is occupied by another Treecreeper species, the Short-toed Treecreeper. Our Treecreeper spread north into Britain at the end of the last ice age, following the advance of conifers, but the Short-toed Treecreeper would not have been able to follow until broad-leaved woodland became established. However, by the time this had happened, Britain had already become separated from the mainland and the Short-toed Treecreeper, being sedentary in nature, only managed to make it as far north as the Channel Islands (where it is sometimes recorded from Garden BirdWatch gardens). The sedentary nature of the Treecreeper has also been quantified, thanks to the hard work of BTO ringers. Research carried out in a Nottinghamshire wood has shown that adult Treecreepers rarely move more than 500m. This suggests that, once they have established a territory, they remain within its boundaries, although these are surprisingly large for such a small bird.

In flight, the narrow tail and wings are striking.

Treecreeper

Short-toed Treecreeper

NESTING

It is amazing to think that a pair of Treecreepers can squeeze their nest in behind a flap of loose bark. Here they put together a base of twigs onto which they add a nest of grass, moss, lichens and wood chips. Finally they line the nest with hair, wool and feathers (adding moss if they have built in a largely coniferous woodland). It takes about a week for the pair to build the nest with both birds involved in the early stages and the female alone adding the lining. With a clutch size of five to six eggs, things can get a little cramped as the resulting youngsters approach the age at which they will fledge from the nest. Even after leaving the young may return to the nest to roost over the next few nights.

When climbing a tree trunk, the stiff tail provides invaluable support.

NEW OPPORTUNITIES

Researchers discovered that Treecreepers have taken to excavating small roosting cavities (about the size of half a boiled egg) in the soft bark of Wellingtonias. This large conifer was first introduced into Britain during the middle of the 1800s and it would have been a number of decades before individual trees reached a size suitable for the Treecreepers to use. The first recorded case of a Treecreeper roosting in this way was noted in the early 1900s, since when the behaviour has spread, such that it is now a common occurrence where tree and bird coincide.

STATUS

Green listed, not of conservation concern.

ATTRACTING TREECREEPERS

One way of attracting Treecreepers is to try rubbing fat or suet into the bark of trees. A specially designed wedge-shaped nestbox is another possibility but the occupancy rates of these are low, suggesting that we have not quite got the design right to match the natural sites they favour under loose bark.

A Treecreeper's plumage is a good match for the tree bark on which it spends much of its life.

SEASONAL TRENDS

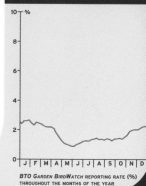

BTO *Garden BirdWatch* reporting rate (%) throughout the months of the year

BTO
BIRDFACTS

FAMILY
Corvidae

LENGTH
34cm

WINGSPAN
55cm

WEIGHT
170g

HABITAT
Woodland.

FOOD
Insects (especially beetles and caterpillars), seeds, fruit, carrion, small vertebrates and eggs.

NEST
A rough base of twigs lined with roots, grass and leaves, placed in crown of tree or tall shrub.

EGGS
4-5 eggs, pale blue-green, olive or sandy buff in colour, finely speckled all over.

EGG SIZE
31.6 × 22.9mm

INCUBATION
18 days. 1 brood.

BREEDING SEASON
April to July

POPULATION SIZE
160,000 breeding territories.

TYPICAL LIFESPAN
Unknown

MAX. RECORDED AGE
16 years 9 months

JAY
Garrulus glandarius

Jays are usually wary birds within the woodland habitats they favour. In gardens they can become surprisingly approachable, most probably because this is a habitat within which they are rarely persecuted.

The white rump is striking in flight.

THE IMPORTANCE OF OAK

Jays are at their most conspicuous during the autumn, when they often travel some distance to find abundant supplies of acorns. Birds will typically collect and hold 3-4 acorns in their adapted gullet (plus one in their bill) before returning to their territories, where the acorns will be stored in the ground. You may be fortunate enough to witness a Jay hoarding acorns in your lawn. First, the bird makes a hole in the ground at a 45-degree angle using its bill. One or two acorns are then deposited in the hole before the bird uses sideways movements of its bill to cover its store. It has been estimated that a single Jay may hoard up to 3,000 acorns during a month and this makes its ability to relocate these stores all the more amazing. Sometimes, especially during the summer, Jays can use the presence of a newly germinated oak sapling to reveal the presence of a hidden acorn attached to its base. When it comes to eating an acorn, a Jay will grasp it in both feet, holding it against the perch, before prising off bits of the shell to make a hole through which it can feed on what is inside. Acorns can be an unpredictable food source and there are years when very few acorns are produced. During such years, substantial numbers of Jays from northern and eastern Europe arrive in Britain in search of food.

IT'S A FACT...

The close association of the Jay with oak is highlighted in its Latin name. The word 'glandarius' comes from the Latin 'glandis' which means 'acorn'.

juvenile

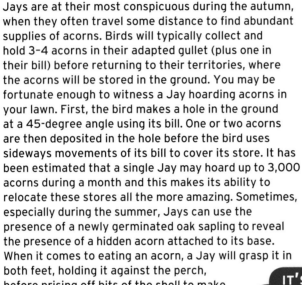

IT'S A FACT...

The stunning azure blue wing feathers are still prized for tying into the flies used for fly-fishing.

THE USE OF GARDENS

Results from both the Garden Bird Feeding Survey and Garden BirdWatch illustrate the use that Jays can make of gardens. In many cases they visit gardens early in the morning to take peanuts, scraps and even marmalade on toast. Some enterprising individuals have learned to up-end hanging peanut feeders in order to empty the contents.

adult

OTHER FOODS

While acorns are the main food in autumn, a wide range of other foods is taken throughout the rest of the year. The Jay's diet can best be described opportunistic, with beetles and caterpillars taken when they are widely available. However, in summer the eggs and young of other birds form an important part of the diet. These provide much-needed protein for the Jay's own chicks, developing in the nest. Carrion of all sorts is also readily eaten and Jays may scavenge from road casualties alongside other scavengers like Magpies and Carrion Crows. While Jays seem quite happy to forage on the ground they are equally at ease collecting caterpillars from the canopy.

SOCIAL BEHAVIOUR

Jays usually do not gather together in large flocks. The small groups that may form during spring are thought to be unpaired birds looking for partners, though some of these may be established pairs on their territory with those young from the previous year that have yet to disperse. Such youngsters are soon nudged on their way come the start of the breeding season proper. Even established pairs tend to spend much of their time apart, seemingly avoiding close contact except when defending their territory or mating.

Adult swallowing an acorn.

PERSECUTION

The Jay is one of just 13 species that can be legally controlled throughout the year, reflecting their habit of taking the young and eggs of various gamebirds. Persecution has helped to shape the distribution of the Jay throughout Britain and Ireland, and has also played a part in setting population levels. The Jay is now more widely distributed than it was in the 19th century and is only really absent from upland areas and the fens.

A VOCAL REPERTOIRE

Although best known for its strident alarm call - a harsh, almost rasping screech - the Jay has quite a repertoire. In fact, the Jay can deliver quite an impressive song. This is composed of a medley of phrases and calls, many of which are mimicked from other sounds heard by the bird in the local area. Some of these can be fairly accurate copies of the original sound, sometimes catching out a human observer.

adult

STATUS

Green listed because, despite fluctuations, stable over the longer term.

SEASONAL TRENDS

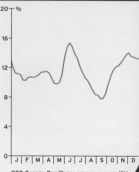

BTO Garden BirdWatch reporting rate (%) throughout the months of the year

BTO BIRDFACTS

MAGPIE
Pica pica

The Magpie is a bold and conspicuous garden visitor. The striking black and white plumage, together with the harsh chattering 'chacha-cha-cha' call, should be familiar to most.

A gathering of Magpies.

FAMILY
Corvidae

LENGTH
45cm

WINGSPAN
56cm

WEIGHT
Male 240g; female 200g

HABITAT
Open woodland, farmland, towns.

FOOD
Catholic in its tastes, taking invertebrates, fruit, seeds, carrion and small vertebrates.

NEST
Loose, bulky nest made of twigs, usually extended to form a dome. The nest is lined with mud and then a layer of fine material. Position usually high in a tree but may be lower.

EGGS
5-6 eggs, very variable in colour from pale blue through to olive with heavy speckling.

EGG SIZE
34.7 × 24mm

INCUBATION
20 days. 1 brood.

BREEDING SEASON
March to June

POPULATION SIZE
590,000 breeding pairs.

TYPICAL LIFESPAN
5 years

MAX. RECORDED AGE
21 years 8 months

ON THE UP
Over the last 50 or so years, Magpies have increasingly colonised suburban and urban habitats, making use of the food we provide. This increase in urban and suburban populations ties in with an increase in the Magpie population across most habitats, following a decline in the amount of persecution taking place within the wider countryside. The distribution of breeding Magpies extends right across England, Wales and Ireland. However, the species is more localised within Scotland, being absent from most upland areas as well as from large parts of the Borders. Magpies obviously find the urban environment an ideal one in which to live. The abundance of food, together with lower levels of persecution, has resulted in breeding densities higher than those seen in farmland habitats. The benefits of living in association with Man can be seen by the establishment of Magpie populations in the middle of the treeless tundra regions of Russia, where the birds nest on the ground or on buildings, alongside newly established human settlements.

BEHAVIOUR
Immature birds gather in non-territory-holding flocks, giving rise to one of the most interesting aspects of Magpie social behaviour, known as ceremonial gatherings. These begin when small groups of immatures target the territory of an established pair. Such incursions elicit a noisy response from the territory owners who chase and attack the intruders, rapidly attracting onlookers in the form of neighbouring pairs and other individuals. These intrusions usually last for a few minutes but they can last for much longer before the birds move on to try their luck elsewhere. Since there is a dominance hierarchy within these groups, it seems likely that it will be the dominant male and his mate that seize the annexed ground if the incursion is successful.

IT'S A FACT...
The image of a Magpie hoarding jewellery or other shiny objects in the nest (or elsewhere) is something of a myth and there is no evidence whatsoever that wild Magpies hoard anything other than food!

A striking black and white pattern is obvious in flight.

From the front, the black chest contrasts with the white belly.

In good light, a blue sheen is seen on some of an adult bird's feathers.

adult

juvenile

A green sheen is seen on the tail in good light.

A CATHOLIC DIET

Much of the success of the Magpie comes from the catholic diet and the adaptability that this species shares with other corvids. Various invertebrates, especially insects, form the bulk of the diet, but fruit, grain, berries, carrion and small mammals are also taken. The eggs and young of other birds form part of the diet during the breeding season. One of the most interesting aspects of Magpie feeding behaviour is that of food hoarding, a habit shared with other species like Coal Tit, Jay and Nuthatch. In the case of the Magpie, the caching of food items is of a more short-term nature, with perishable items often recovered and eaten within a few days of being stored. Items are typically deposited in a hole in the ground, dug using the bill, and into which food is regurgitated from a small pouch under the tongue.

THE MAGPIE'S NEST

The nest is often quite obvious, large in size and placed in the fork of a tree or large shrub. The fact that nest building is a noisy affair, with both birds of a pair chattering, further adds to the ease with which nests can be found. Nest building may begin as early as February but eggs are not normally laid until April. Research carried out in Sheffield has shown that older birds are more likely to finish the nest off with a dome than those birds building a nest for the first time. Inside, the nest is held together with a layer of mud and an inner layer of grass.

ASYNCHRONOUS HATCHING

Magpies begin to incubate their eggs before they have completed the entire clutch. This results in some eggs hatching earlier than others, a process known as asynchronous hatching. It is the smallest chicks that will die if there is insufficient food available for all the youngsters in a nest, thereby helping to ensure that at least some survive.

MAGPIES AND SONGBIRDS

The perceived 'problem' of Magpie predation on songbirds has its roots in the fact that, while Magpie populations have been recovering from former levels of persecution, the populations of songbirds have declined. Alongside this, instances of Magpie predation of eggs and chicks have tended to elicit a strong emotional response from observers. Such a response to nature 'red in tooth and claw' is understandable, but suggestions that Magpie predation is the cause of widespread population declines are unsupported by scientific evidence.

STATUS

Steady increase up to 1980s; numbers have since stabilised or even declined slightly. Green listed.

SEASONAL TRENDS

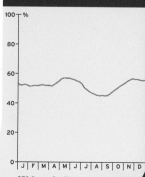

BTO GARDEN BIRDWATCH REPORTING RATE (%) THROUGHOUT THE MONTHS OF THE YEAR

BTO BIRDFACTS

JACKDAW
Corvus monedula

FAMILY	Corvidae
LENGTH	34cm
WINGSPAN	70cm
WEIGHT	220g
HABITAT	Farmland, open woodland and urban areas.
FOOD	Largely composed of invertebrates, fruit and seeds but also carrion, small vertebrates and eggs.
NEST	A messy base of sticks, often with lumps of mud or dung, lined with wool, hair, rotten wood and moss.
EGGS	4-5 eggs, pale greenish-blue with darker blotches and spots.
EGG SIZE	35.7 × 25.5mm
INCUBATION	20 days. 1 brood.
BREEDING SEASON	April to June
POPULATION SIZE	503,000 breeding pairs.
TYPICAL LIFESPAN	5 years
MAX. RECORDED AGE	15 years 11 months

Although Jackdaws occur in greatest numbers in areas of mixed farming and pasture, well-stocked with cattle and sheep, they are equally at home in large gardens, where they will often nest in chimneys, tree cavities or even nestboxes.

SOCIAL LIFE

Jackdaws are gregarious birds, nesting colonially and gathering together with other crows in large winter roosts. However, the basic unit of Jackdaw society is the pair and individual birds are extremely faithful to their mates. The pair bond is so strong that birds will remain together even when they have been unsuccessful in their breeding attempts over several seasons. One of the reasons suggested for the presence of such a strong pair bond is that the birds have to put in a huge amount of effort early in the breeding season in order to find enough food for their developing chicks. Jackdaw eggs hatch asynchronously (incubation begins before all the eggs have been laid) and this means that the young differ in age. The oldest chicks can hold their gaping mouths higher than their siblings and so get the lion's share of the food being delivered to the nest. It is only during those years when food is really abundant that the parents manage to rear all the chicks that hatch.

FOODS AND FEEDING

Although Jackdaws are catholic in their diet, they concentrate on invertebrates during the breeding season, taking the larvae of moths and flies from the surface of the ground. Bird tables are probably more frequently visited than many people imagine, Jackdaws often arriving very early in the morning to take whatever is available, especially during early summer.

adult

Jackdaws often feed alongside Rooks.

Where they are not discouraged, Jackdaws are regular and bold visitors to the bird table, consuming large amounts of food.

NESTING BEHAVIOUR

Most Jackdaw nests are in cavities (including chimneys) and the male will drop sticks into the hole until some begin to catch, allowing a platform to be constructed. Once this has been completed, the female will create a bowl and add a lining, often of wool. Prior to nest building, the pair will often perch at the entrance, a clear signal that the site will be used in the weeks ahead. Once the eggs hatch, the female will remain with the chicks over the first 10-12 days, while the male brings food for both her and the chicks. At this stage, the female may leave the nest for brief periods, to preen and exercise, leaving the male to guard the young. The chicks remain in the nest for almost five weeks and, even once they have fledged, the adults still have to provide food for them for several more weeks.

IT'S A FACT...

Jackdaws tend to fill their nesting cavities with sticks, the bigger the cavity the more sticks that are used to fill it; this can be a problem if the cavity in question happens to be your chimney.

adult

LEGENDARY INTELLIGENCE

The intelligence of Jackdaws is legendary and you only need to watch them figuring out how to exploit a new feeding opportunity to realise just how clever they are. Although they are reputed to steal jewellery like Magpies there seems to be no truth in this, at least for genuine wild birds. On the other hand, tame birds can be taught to perform tricks. One particular individual was trained by a group of Italian thieves to swoop down and grab money from users of a cash machine. This intelligence may extend to the care of young and other relatives. A fully-grown young Jackdaw, trapped in a nest by a tangle of wool, was fed right through the summer by what was presumably one of its parents. The bird was then freed by a helpful Garden Birdwatcher after something like 12 weeks in the nest.

IT'S A FACT...

Ever resourceful, Jackdaws have been reported hawking flying ants alongside gulls and stealing fish from Puffins.

adult

STATUS

Population has increased since 1960s, hence Green listing.

IDENTIFICATION

The Jackdaw is the smallest of our black crows, being about two-thirds of the size of a Rook or Carrion Crow. Young Jackdaws lack the distinctive grey colour to the nape and sides of the head of the adult but do have wonderful blue-grey eyes. As the chicks get older, the eye changes colour, becoming brown in their first winter and pearly white when they attain the adult breeding plumage.

adult

SEASONAL TRENDS

BTO GARDEN BIRDWATCH REPORTING RATE (%) THROUGHOUT THE MONTHS OF THE YEAR

CARRION & HOODED CROWS
Corvus corone & Corvus cornix

BTO BIRDFACTS

FAMILY	Corvidae
LENGTH	46cm
WINGSPAN	98cm
WEIGHT	510g
HABITAT	Woodland edge, farmland, upland fringe and urban areas.
FOOD	Wide range of invertebrates but also grain, carrion, eggs and scraps.
NEST	Several layers of twigs, turf, smaller twigs and lining of softer material. Nest itself is placed high in tree, sometimes on a ledge or the ground.
EGGS	3-4 eggs. Light-blue to green with darker speckles and spots. (3-6 eggs).
EGG SIZE	43 × 30mm
INCUBATION	18-20 days. 1 brood. (18-19 days.)
BREEDING SEASON	March to July
POPULATION SIZE	790,000 breeding territories. (160,000 breeding territories.)
TYPICAL LIFESPAN	4 years
MAX. RECORDED AGE	21 years 9 months (16 years 9 months)

NOTE Where different, Hooded Crow given in brackets,

Both of these crows may visit gardens: the Hooded Crow to rural sites in Ireland and the northwest of Scotland; the Carrion Crow to a range of garden types across much of Britain. Late spring and early summer, when the birds are breeding, is the period when they are most often seen using gardens.

CHANGING FORTUNES

The range of the Hooded Crow has retreated northwards over the years, as the Carrion Crow has expanded its range. Coincident with this retreat has been a decline in the number of Hooded Crows reported wintering in eastern England. These winter visitors were most likely of Scandinavian origin, as suggested by recoveries here of individuals ringed in Norway and Sweden. Birds from these populations now winter further north than they used to, a sign of climate change, and so they no longer reach our shores.

DIETARY DISAGREEMENTS

The highly adaptable nature of both crow species has brought them into conflict with Man. Gamekeepers often control crows because they will take eggs and young gamebird chicks. Similarly, sheep farmers have controlled them because they scavenge carrion and have been blamed for killing young lambs. Insects are the staple diet in summer, supplemented by carrion and other scavenged food, while in winter grain is important. Individuals have also been recorded opening milk bottle tops to get at the cream and taking shellfish on the seashore. Carrion Crows are expert nest finders and predate the nests of songbirds, especially in gardens where there are few places for open-nesting species to hide their nests. While the more conspicuous Magpies tend to get the blame for much of this predation, crows tend to conduct their raids early in the day, often just after dawn, and so escape unnoticed. Most of this nest predation takes place during the short period when the crows have young of their own in the nest. Since they only produce one brood a year, most of the main prey species are able to raise other broods successfully either earlier or later in the year.

IDENTIFICATION

The Carrion Crow has a heavier black bill than the Rook and lacks the shaggy thighs. It also has a flatter crown to the head. Both Rook and Carrion Crow are noticeably smaller than the Raven. The Hooded Crow has a black head, wings and tail and a grey body. The typical call of these crows is a resonant 'kraa', stronger than the rather flat call of the Rook. Carrion and Hooded Crows are usually seen in pairs or small groups.

Carrion Crow

Rook

Raven

SOCIAL BEHAVIOUR

Unlike Rooks and Jackdaws, both crow species are solitary nesters, defending large territories around the nest site for much of the year. The nest is usually built high in a tree but may be on a cliff face, building or other structure. Territory boundaries are less aggressively defended against neighbouring pairs than against other birds and this allows for mutual defence against intruders. Flocks usually consist of immature birds and birds of breeding age lacking a territory. Communal roosts form in the winter, largely involving birds from these floating flocks, and sometimes involving other species like Rook and Jackdaw.

Hooded Crows mobbing a Buzzard.

Hooded Crow

Hooded Crow

FAR LEFT:
Hooded Crow
LEFT:
Carrion Crow

Not currently listed.

Carrion Crow

Hooded Crow

TWO DIFFERENT BIRDS

Traditionally, the Carrion Crow and the Hooded Crow were regarded as being two races of the same species, since they interbreed freely and produce fertile offspring. However, hybrids produced from matings between birds of the two forms are very unsuccessful, resulting in a striking lack of mixed pairs. A very narrow 'hybrid zone', where intermediate forms occur, is a consequence of this, reducing the amount of gene flow. Other research has highlighted subtle differences in the vocalisations made by the two forms and in habitat use. Taken together, these findings have enabled taxonomists to recommend that the two forms be treated as separate species on the British List. As such, the two were 'split' in 2002. The Carrion Crow is found across England and Wales, together with southern and eastern Scotland, while the Hooded Crow is largely restricted to the Highlands, Northern Isles and Outer Hebrides of Scotland and also to Ireland.

Carrion Crow

Hooded Crow

IT'S A FACT...

Hooded Crows have been spotted catching Salmon smolt at a Scottish Salmon trap on the River Meig.

Combined trends for Carrion and Hooded Crows.

BTO *Garden BirdWatch* REPORTING RATE (%)
THROUGHOUT THE MONTHS OF THE YEAR

BTO BIRDFACTS

ROOK
Corvus frugilegus

FAMILY	Corvidae
LENGTH	45cm
WINGSPAN	90cm
WEIGHT	310g
HABITAT	Pastoral farmland.
FOOD	Invertebrates, especially those taken from the soil; carrion, cereal grain and small vertebrates.
NEST	A base of sticks, lined with grass, moss, feathers and leaves. Placed high in tree canopy.
EGGS	3-4 eggs, light blue to dull green in colour and covered with dark olive streaks.
EGG SIZE	40 × 28.3mm
INCUBATION	15-17 days. 1 brood.
BREEDING SEASON	March to June
POPULATION SIZE	1.2 million breeding pairs.
TYPICAL LIFESPAN	5 years
MAX. RECORDED AGE	22 years 11 months

IT'S A FACT...

A gathering of Rooks is commonly known as a 'parliament', quite a fitting description.

Rooks are closely associated with agricultural land below 300m and it is not surprising that the majority of Garden BirdWatch records come from large rural gardens. However, in some parts of their range, Rooks will readily enter the outer suburbs of large towns in search of food or nesting opportunities.

Rookery in spring.

Adult showing glossy sheen to plumage.

DIET

Rooks are less catholic in their diet than the other large crows and specialise on soil-living invertebrates, particularly earthworms, and cereal grain. Birds feed by probing and digging with their long, pointed bills. Feeding Rooks have a confident upright stance as they walk about in search of suitable prey. Recent changes in farming methods have caused problems for Rooks and have brought them into conflict with farmers. Changes in cropping practices mean that germinating cereal crops are now available to Rooks at the very time of year when other foods are difficult to find. Being adaptable birds, and unlikely to miss a free lunch, the Rooks can inflict localised, but heavy damage, to these crops. In response, some landowners shoot young Rooks on the nest in an attempt to control the population. Controlling a population by removing a proportion of the young birds is often unsuccessful. Many young birds would die anyway and the remaining birds end up with a greater share of the available resources, which means that they are more likely to survive than they would otherwise.

ROOKERIES

The colonial nests made by Rooks, placed high in mature trees, are familiar to most as 'rookeries'. These are noisy affairs, especially during the early part of the breeding season, when pairs reclaim nests used the previous year. Most English rookeries contain fewer than 50 nests and only a handful have been recorded with more than 300 nests. A few really big rookeries have been recorded in the past, notably on large Scottish estates where, for example, some 2,274 nests were noted at the Haddo House rookery near Aberdeen in the 1970s. The trend, certainly within England, seems to be towards smaller and smaller rookeries, perhaps a reflection of food availability, loss of suitable nesting trees or increased levels of disturbance. During the period of nest construction, raiding parties, composed primarily of young birds, attempt to steal twigs from unguarded nests. For this reason, breeding females will typically guard the nest while their mate seeks out suitable twigs. Only when the lining of the nest begins is the female confident enough to join the male in collecting material. Pairs can remain together for several years, if successful in their breeding attempts, and roost together even when they join the huge winter roosts.

A CONFIDING BIRD?

The persecution of Rooks in the wider countryside has made them wary of associating with humans. However, in some urban areas, or in large rural gardens where they are regularly fed, they may become remarkably confiding and approachable, to the extent where they become a nuisance, monopolising feeding stations.

WINTER ROOSTS

Winter roosts often contain many thousands of individuals from different rookeries, including some that may have travelled up to 20km. One roost in Scotland was estimated to contain some 65,000 birds - a staggering figure. In order to reach the roosts, small groups of Rooks will travel along well-defined lines of flight, arriving close to the roost before dark and joining other birds that have already ceased feeding and are preening or generally loafing around. As it gets dark, the birds move into the trees that form the roost site.

IDENTIFICATION

The three larger species of black 'crows' can prove problematic when it comes to identification. However, during the breeding season, adult Rooks show an off-white base to the bill, steep forehead and pointed bill, which tapers along its length (compare this with that of Carrion Crow). The feathers on the flanks appear loose, especially when the bird is striding across the ground. Rooks utter a range of calls but the most familiar is the rather flat-sounding 'kaah'. Both Rook and Carrion Crow are noticeably smaller than the Raven, the largest of our crows and a rare garden visitor.

Long primaries allow for acrobatic flight.

ABOVE: **Pair bonds are reinforced with affectionate-looking interactions.**

juvenile

Despite their size, Rooks are surprisingly agile visitors to the bird table.

IT'S A FACT...

The Rook's plumage is not simply black, but is instead dressed with a wonderful mix of blues and purples, reminiscent of the way in which oil on water separates out into a kaleidoscope of colours.

STATUS

Green listed.

SEASONAL TRENDS

BTO GARDEN BIRDWATCH REPORTING RATE (%) THROUGHOUT THE MONTHS OF THE YEAR

BTO BIRDFACTS

STARLING
Sturnus vulgaris

winter adult

FAMILY	Sturnidae
LENGTH	22cm
WINGSPAN	40cm
WEIGHT	78g
HABITAT	Almost anywhere that provides short grass for feeding.
FOOD	Mostly insects and other invertebrates in breeding season; seeds, fruits and bird-table scraps at other seasons.
NEST	Hole in tree, building or cliff, or a nestbox with 45mm diameter hole.
EGGS	4-7 eggs, shades of pale blue in colour.
EGG SIZE	30 × 21mm
INCUBATION	12-14 days. 1-2 broods
BREEDING SEASON	March to July
POPULATION SIZE	8 to 10 million in summer; numbers boosted by migrants in winter.
TYPICAL LIFESPAN	5 years
MAX. RECORDED AGE	17 years 7 months

With a somewhat unfair reputation of being an argumentative and greedy bird, the Starling is a common, though declining, visitor to most gardens. Starling numbers are swelled in the winter by the arrival of birds from elsewhere in Europe, many performing spectacular aerial displays before entering their nightly roosts.

OUR STARLING POPULATIONS

Although it is easy to see Starlings throughout the year, there are actually two different components to our Starling population. First, there is a resident breeding population, which also remains here for the winter. This is joined in the autumn by an additional wintering population, drawn from breeding populations elsewhere in Europe, which follow an autumn migration fly-way that runs east to west across Europe. Birds from The Netherlands arrive first, followed by birds from Germany and Scandinavia. The last birds to arrive come from Poland and Russia. The Garden BirdWatch reporting rate increases dramatically in the autumn, highlighting the arrival of these wintering birds. Monitoring work by the BTO has highlighted a long-term decline in Starling populations, both here in Britain and elsewhere in Europe. Such declines are thought to be linked to a decline in feeding opportunities, reflecting a general loss of permanent pasture (the preferred foraging habitat) and increased stocking densities of livestock on the areas that remain.

FOODS AND FEEDING

During the breeding season, Starlings feed predominantly on invertebrates, especially leatherjackets (the larvae of craneflies), which are taken from short grassland – such as lawns. In late summer and autumn, the amount of plant material taken increases, reflecting a change in food availability. This seasonal shift in diet is matched by a change in the length of the Starling's intestine, which gets longer to cope with a diet composed largely of plant material; plant material is harder to digest than animal material. The Starling shows other adaptations linked to its way of feeding. Modifications to the skull and the musculature around it, allow the Starling to push its bill into the soil and then open it to create a hole. As the bill opens, the eyes rotate forwards to give the bird binocular vision, useful for spotting prey within range of the bill.

autumn bird

IT'S A FACT...

A Starling's body shape may reflect it's mood at the time; sleeked feathers (*right*) may indicate a sense of alertness or agitation, while fluffed-up birds (*far right*) are typically more relaxed.

winter adults

NESTING REQUIREMENTS

Breeding Starlings have two main requirements: a suitable hole in which they can nest and nearby grassland where they can feed. Such requirements are met in many habitats, including urban ones, so that the Starling has become an abundant species across much of Britain and Ireland. It may seem surprising therefore, that just 150 years ago the Starling was a scarce breeding bird. A wide range of holes is used, including those under the eaves of houses but these are becoming less common thanks to the increasing use of uPVC bargeboards. Starlings will use nestboxes and have even been recorded taking over the nests of House Sparrows and Swifts. Mature males may establish breeding territories as early as January, each of which is a small area around the selected nesting hole. Many males try to defend several holes in the hope of attracting a mate to each one, but competition for nest sites is such that most end up with just the one hole. Females can be equally opportunistic when it comes to breeding. It is quite common for a female Starling to deposit one of her eggs in another female's nest – a behaviour known as brood parasitism. To reduce the chances of her rogue egg being spotted, she may remove one of the nest owner's eggs after depositing her own, so that the numbers still add up. It is these eggs that are sometimes to be found on the ground.

summer male

STARLING PLUMAGE AND AGEING

Starling plumage changes as the bird matures from juvenile to adult. Adult plumage also changes at different times of the year.

FLOCKS

During the winter months, Starlings may gather in vast roosting flocks, performing amazing aerobatic displays before dropping into favoured roost sites. Many of these sites are in the countryside but those in urban areas may cause a nuisance, by contaminating buildings with their droppings.

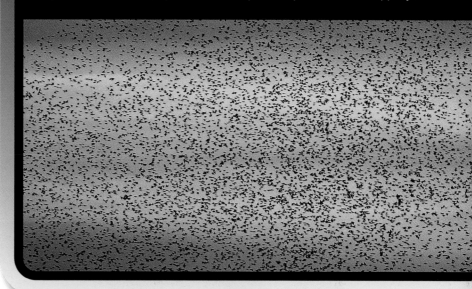

UNFAIR REPUTATION

The reputation for being greedy and argumentative is a little harsh on the Starling. After all, this is a highly sociable species with a fascinating series of behaviours that have evolved from communal living and feeding. Those garden birdwatchers who regard the Starling as a pest, should take a closer look at this delightful bird.

STATUS

Widespread but declining, hence Red listed.

SEASONAL TRENDS

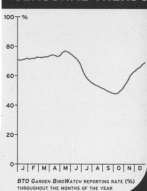

BTO GARDEN BIRDWATCH REPORTING RATE (%) THROUGHOUT THE MONTHS OF THE YEAR

| juvenile | immature first autumn | immature first winter | adult summer | adult winter |

BTO BIRDFACTS

HOUSE SPARROW
Passer domesticus

FAMILY
Passeridae

LENGTH
14cm

WINGSPAN
24cm

WEIGHT
34g

HABITAT
Tends to be found breeding alongside Man, either around farms or areas of habitation.

FOOD
Mainly large seeds but also takes berries and scraps. The chicks are fed predominantly on insects.

NEST
Usually in a hole, though may be freestanding. Made from grass or straw and lined with softer material.

EGGS
4-5 eggs, white or faintly grey, covered with darker blotches.

EGG SIZE
22.5 × 15.7mm

INCUBATION
13-15 days. 2-3, sometimes 4, broods.

BREEDING SEASON
March to August

POPULATION SIZE
6 million breeding pairs.

TYPICAL LIFESPAN
3 years

MAX. RECORDED AGE
12 years

The House Sparrow is a familiar visitor to many garden feeding stations, favouring those where there is suitable cover close to bird tables and hanging feeders. Populations have declined in many areas, the population roughly halving in size over recent decades.

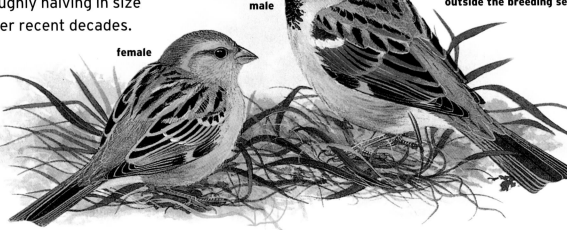

female

male

House Sparrow pairs sometimes maintain a loose association outside the breeding season.

A SOCIABLE BIRD

The communal nature of the House Sparrow is at its most obvious during the winter, when small flocks often gather in trees or bushes and 'chatter' between bouts of feeding. House Sparrows are most attracted to hanging seed and peanut feeders that have plenty of space for several birds to feed at once, and are within easy reach of cover. House Sparrows are seed-eaters by nature but within the garden environment many different kitchen scraps are taken. Winter-feeding flocks usually contain birds from several local colonies and it is through these that the young disperse away from the colony in which they were born. Many of the individuals in these flocks will roost together in dense vegetation, although birds that have bred during the previous summer will often roost in the nest site.

NESTING

House Sparrows establish their nest sites and form pairs early in the season. In addition to nesting in cavities, House Sparrows will also build rather untidy nests within dense vegetation. House Sparrows sometimes take over nests built by House Martins, and have even been recorded living and breeding underground in a coal mine.

male

female

IN DECLINE !

Dramatic declines in House Sparrow populations have been revealed by BTO monitoring programmes. The current population is thought to number between 6 and 7 million pairs, having fallen from around 12 million pairs during the 1970s. Just under two-thirds of these occur in towns and villages, including about 2 million pairs in suburban habitats, mostly in southern, eastern and central England. Results from the BTO's Garden Bird Feeding Survey show that the decline began in the mid-1980s in rural gardens and slightly later in suburban gardens. Interestingly, the rate of decline in London and southeast England has been far more pronounced than in Scotland or Wales. There is evidence to suggest that the decline has been most acute in urban and rural areas and less pronounced in suburban habitats. This supports the theory of some Dutch scientists, who believe that suburban gardens are the preferred habitat of House Sparrows and act as source populations for poorer quality habitats in rural and urban areas. Results from the BTO House Sparrow Survey tend to support this idea, having demonstrated the importance of houses with gardens, as opposed to high density housing with little or no space for gardens.

A BADGE OF HONOUR

Established birds pair for life and tend to use the same nest site each year. Males will perch above the nest hole and proclaim their ownership. Young birds will attempt to attract a mate once they have found a hole and the females use the black bib of the male as a means of assessing his suitability. Dominant males have larger bibs, so an established older male that has lost its mate will probably secure a new female before a younger, less dominant individual.

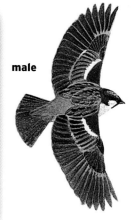

male

female

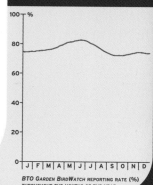

male

female

WHAT'S BEHIND THE DECLINE

The underlying causes of the declines seen across rural, suburban and urban habitats may well be different, since House Sparrows rarely move very far. Within rural areas, changes in farming practices, notably the loss of overwinter stubbles and suitable nest sites, may have driven the decline. In urban and suburban habitats a wider range of factors may be implicated. These include: (1) loss of nest sites as new designs of roof tile and barge board restrict access to traditional nesting sites, (2) increased levels of predation from Sparrowhawks and domestic cats – one study in a Bedfordshire village showed that cats were responsible for 30% of the House Sparrow mortality, (3) increased competition from increasing populations of Feral Pigeons, Collared Doves and Woodpigeons, (4) a decline in sources of weed seeds due to the loss of overgrown, unused, city sites, (5) a decline in availability of those insect species important in chick diet because of increased pesticide use in gardens, (6) increased levels of pollution, (7) changes in the suitability of foods provided at garden feeding stations, (8) loss of suitable roosting sites and (9) increased levels of disease transmission.

IT'S A FACT...

House Sparrows are versatile feeders but grain is by far the preferred food. As such, they thrive on cheap seed mixes that tend to have a lot of grain in them.

STATUS

Red listed because of pronounced decline in breeding numbers.

SEASONAL TRENDS

BTO Garden BirdWatch REPORTING RATE (%) THROUGHOUT THE MONTHS OF THE YEAR

BTO BIRDFACTS

FAMILY	Passeridae
LENGTH	14cm
WINGSPAN	21cm
WEIGHT	24g
HABITAT	Farmland.
FOOD	Mostly seeds of various kinds, though feeds its young on insects.
NEST	Usually in a hole, though may be freestanding. Made from grass or straw and lined with softer material.
EGGS	5-6 eggs, heavily speckled with brown on a white background.
EGG SIZE	19.3 × 14mm
INCUBATION	12-13 days. 2-3 broods.
BREEDING SEASON	April to August
POPULATION SIZE	68,000 breeding territories.
TYPICAL LIFESPAN	2 years
MAX. RECORDED AGE	10 years 10 months

TREE SPARROW
Passer montanus

The Tree Sparrow is the less-familiar relative of the House Sparrow, traditionally found in farmland and rural gardens around the edges of villages across much of England, Wales and lowland Scotland. In Ireland, the species is largely restricted to coastal localities, though inland colonies have become established in Northern Ireland.

For a mainly brown bird, the Tree Sparrow is striking and well-marked.

IDENTIFICATION

Both male and female Tree Sparrows are of similar appearance and can be distinguished from House Sparrows by differences in the colour of the plumage on and around the head. Tree Sparrows have a characteristic warm red-brown crown, white patches to the sides of the head and a small black cheek patch. There is also a narrow white collar. The black bib is much smaller than that seen in a male House Sparrow, reaching only the top of the chest and narrow in outline. Juvenile birds are similar in appearance to the adults but are duller in colour and have dark rather than white cheeks. They still have the warm red-brown cap. The calls are rather similar to those uttered by the House Sparrow but sharper in tone, with a distinctive 'tcheck' or 'tchup'.

FOODS AND FEEDING

While insects are included in the diet during the breeding season, and the young are raised entirely on them, Tree Sparrows specialise on the seeds of arable weeds like chickweed. Gardens are more often visited in the winter months, when birds take mixed seed, fat and peanuts, preferably from the ground, but also from bird tables and hanging feeders. Visiting flocks often indulge in social singing, a behaviour that is thought to reinforce the cohesion of the flock.

NESTBOX LIVING

Breeding Tree Sparrows can be attracted to nestboxes if you have a suitable rural garden. Such boxes should have an entrance hole of 28mm diameter or larger. Because they are semi-colonial nesters you should find that a group of half a dozen together is more attractive than a box on its own. If you are fortunate enough to attract Tree Sparrows into your nestbox make sure that you leave them alone. Tree Sparrows will desert nests much more readily than tits.

Adult perched near its nest under a roof tile.

IT'S A FACT...

Males Tree Sparrows perch at the nest site during spring to proclaim ownership and, if unpaired, to attract a mate.

CHANGING FORTUNES

The Tree Sparrow population has shown both dramatic increases and decreases over the last 100 years, with a fivefold increase in the 1960s and 1970s followed by an equally large and rapid decline since the 1970s. Although widespread across lowland Britain at the end of the 19th century, its population has been in decline since the 1930s. A slight recovery during the 1960s was linked to an influx of immigrants from the continent but since then things have taken a turn for the worse. Information from the BTO's monitoring programmes suggests a decline of 94% over the last 25 years and the species is now Red listed, as a bird of high conservation concern. As with other farmland seed-eaters, the decline of the Tree Sparrow has been linked to changing farming practices. The loss of overwinter stubbles may have reduced food availability and contributed to a potential fall in overwinter survival. However, the species may also be facing problems during the breeding season. Feeding its young on invertebrates, rather than seeds, may leave the species open to a decline in invertebrate numbers.

Tree Sparrow pair.

Adult in flight.

BREEDING BEHAVIOUR

Tree sparrows breed quite late in the year, with most initiating egg-laying during May. Individual pairs may breed through into August and beyond, attempting two or three broods through the season. Breeding occurs in loose colonies of 10–50 pairs, which fluctuate dramatically. Some build up over a very short period, remain for a few years and then suddenly disappear. Part of the reason for such changes may be colony structure and the availability of nesting and feeding opportunities. Just over one-third of the birds in a typical colony are adults from the previous year, a further 5% are juveniles born at the colony, while the remainder are juveniles from other colonies. Once established, colonies do not appear to increase in size, the surplus juveniles emigrating to join other colonies. It may be this feature, coupled with the short lifespan of Tree Sparrows, that drives the establishment of new colonies and the loss of old ones. Breeding Tree Sparrows tend to pair for life, though only a small proportion of individuals survive through to their second breeding season.

IT'S A FACT...

Some rural Garden Birdwatchers will have Tree Sparrows nesting in their gardens, sometimes in cavities in the roof, in thatch or in old tree holes or nestboxes, but others will have them visiting during the winter months, when garden feeding stations may be an important source of food. Curiously, elsewhere in the world Tree Sparrows are urban birds, often living alongside House Sparrows.

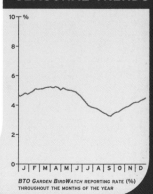

Adult visiting a garden pond to bathe and drink.

STATUS

Red listed because of massive decline in breeding population.

SEASONAL TRENDS

BTO GARDEN BIRDWATCH REPORTING RATE (%) THROUGHOUT THE MONTHS OF THE YEAR

BTO BIRDFACTS

TOP: **male**
ABOVE: **female**

male

CHAFFINCH
Fringilla coelebs

FAMILY	Fringillidae
LENGTH	14cm
WINGSPAN	26cm
WEIGHT	24g
HABITAT	Woodland, farmland and gardens.
FOOD	Insects and spiders are taken during the summer but at other times of the year seeds are favoured.
NEST	A neat, finely woven nest of grass, hair, lichen and rootlets, with a deep cup. Usually placed in fork of tree or bush.
EGGS	4-5 eggs, variable in colour from blue-green to red-brown, with darker streaks and scrawls.
EGG SIZE	19.5 × 14.6mm
INCUBATION	12-13 days. 1 brood.
BREEDING SEASON	April to July
POPULATION SIZE	5.6 million breeding territories.
TYPICAL LIFESPAN	3 years
MAX. RECORDED AGE	11 years 7 months

The Chaffinch is one of the most abundant and widespread bird species to be found in Britain and Ireland. Although mature deciduous woodland supports the highest densities of breeding Chaffinches, they are equally at home nesting in parks and larger gardens, providing that suitable trees and shrubs are present.

FEEDING BEHAVIOUR

Chaffinches are dependent upon tree and shrub cover for nesting and feeding and this may be the reason why the species is virtually absent during the summer from upland areas, from Shetland and from much of the Outer Hebrides. At this time of the year, the adults take small caterpillars and flies from foliage for their growing chicks, as well as feeding upon other insects taken from the ground. During the autumn and winter, the Chaffinch switches from this diet of insects to feed on a wide range of seeds (more than 100 different types of seed have been recorded in the diet). These seeds are taken from the ground, usually from disturbed soil where they have been brought to the surface. Such ground feeding is a behaviour shared with the closely related Brambling, though the two species avoid direct competition with each other by feeding on different sizes of seeds - the Brambling has a bigger and more robust bill than that of a Chaffinch. During the breeding season, resident Chaffinch pairs spend most of their time foraging within their breeding territories but they will feed on other open ground, undefended by neighbouring individuals, if feeding conditions are suitable. These resident individuals usually remain within, or close to, their territories during the winter months as well, especially in central and southern Britain. However, in Scotland (and probably in other northern areas) many may leave their territories and form large flocks, to make use of feeding opportunities elsewhere.

IT'S A FACT...

Chaffinches feed on a range of tree seeds, especially Beech mast. In years with a poor crop of mast they may move further in search of supplies.

RACES

The Chaffinch has evolved into a number of distinct races, the one resident in Britain and Ireland being slightly smaller and more brightly coloured than the continental immigrants. Several islands support their own distinct races, including the Canaries, where the males have a blue-grey back (as opposed to the brown back of our males) that matches the colour on the head.

juvenile

male, March

male, May

male, September

male, November

IT'S A FACT...

The beautiful blue-grey crown of a male Chaffinch is not the result of new feathers moulting through in spring. Instead it is formed as the brown tips of feathers grown the previous autumn wear away over winter, revealing the colour below.

Adult male in flight.

male

WARTS AND ALL

Chaffinches are susceptible to a disease called *Fringilla papillomavirus* and in a recent Dutch study looking at 25,000 Chaffinches, some 1.3% were found to be affected by this wart-forming disease. Cases usually occur in clusters and high proportions of local populations can be affected during outbreaks. The growths occur on the foot or bare part of the leg. They grow slowly and can last for many months, ranging in size from a small nodule up to a deeply-fissured mass that almost engulfs the whole lower leg and foot.

female

male

Female, approaching nest under roof eaves.

ON THE MOVE

During the winter, our Chaffinch population is effectively doubled by the arrival of immigrants from continental Europe. Most of these will have come from migratory populations in Norway, Finland and Sweden, involving birds that have travelled through Denmark, northern Germany, Belgium and northeast France to reach Britain by crossing the English Channel. Many of these visitors will remain to winter within southern and central Britain but some, mostly females, subsequently move further north and west to reach Ireland. The largest winter flocks are usually composed of birds of continental origin, while smaller groups of birds are likely to be British and Irish breeders. One of the most interesting aspects of Chaffinch migration is the differential migration of males and females, with the females migrating further than the males. So extreme is the difference between the two sexes, that females predominate in the wintering population in Ireland, while males dominate populations wintering in Britain, Belgium and The Netherlands. The well-known Swedish naturalist Linnaeus named the Chaffinch 'coelebs' (Latin for bachelor) because the few Chaffinches remaining to winter in Sweden were nearly always males, seemingly abandoned by their mates. Many of the winter immigrants to Britain and Ireland are still present in early spring, when the resident males have attained their breeding plumage and started to sing for a mate.

STATUS

Green listed.

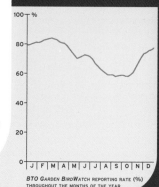

SEASONAL TRENDS

BTO Garden BirdWatch reporting rate (%) THROUGHOUT THE MONTHS OF THE YEAR

BTO BIRDFACTS

BRAMBLING
Fringilla montifringilla

FAMILY	Fringillidae
LENGTH	14cm
WINGSPAN	26cm
WEIGHT	24g
HABITAT	Breeds in birch-conifer woodland but winters here in Beech woodland and may visit gardens.
FOOD	Seeds and berries, with beetles and caterpillars important during the breeding season.
NEST	A cup of moss, lichen and grass, lined with feathers, hair and moss. Usually placed fairly high in a tree.
EGGS	5-7 eggs, light blue to olive-brown in colour with rusty spots.
EGG SIZE	19.5 × 14.6mm
INCUBATION	11-12 days. 1, sometimes 2, broods.
BREEDING SEASON	April to June
POPULATION SIZE	Does not breed here.
TYPICAL LIFESPAN	Unknown
MAX. RECORDED AGE	8 years 7 months

Although the Brambling has been recorded breeding in Britain, it is really a winter visitor, often seen in the company of the closely related Chaffinch. In fact, the Brambling can be considered as being the northern counterpart of the Chaffinch, occupying breeding grounds in the northern forests of Finland, Norway, Sweden and Russia.

IT'S A FACT...

Thomas Bewick, the naturalist and long considered the finest of English wood engravers, famously noted that the Brambling made better eating than the Chaffinch, referring to the bird as the 'mountain finch' (hence 'montifringilla' for the Latin name).

BREEDING ECOLOGY

Within its favoured breeding habitat of birch and mixed birch woodland, the Brambling is usually the second most common breeding species (behind the ubiquitous Willow Warbler). Interestingly, breeding numbers in a given area can vary markedly between years. In those years when spring is late, many Bramblings curtail their northward migration. As a consequence they end up breeding further to the south than they would normally do. Perhaps these late springs are behind the occasional breeding records noted from Britain. The first of these was noted in 1920, when a nest containing seven eggs was found in Sutherland. No further breeding records were confirmed until 1979, when another breeding attempt was noted in Scotland. Since then there have been periodic reports, though some of these have just involved singing males, without additional confirmation that breeding actually took place.

IDENTIFICATION

Bramblings of all ages show a white rump (lacking in the Chaffinch) and this, together with the lack of any white edges to the tail, makes a very useful identification feature. The orange-brown shoulders can be seen on both sexes but are more pronounced in the male. Young males often show a series of black splodges within the orange-brown shoulder. The glossy blue-black colour on the head of the adult male is only visible when in breeding plumage. During the winter, buff tips to the crown feathers obscure the black beneath, which only shows through as the weaker light-coloured tips wear away. Female Bramblings have a grey-brown head but are still more brightly coloured than a female Chaffinch. Bramblings tend to be more nervous when feeding on the ground than Chaffinches and often feed in small flocks. These flocks can be quite noisy, with individuals making a series of nasal sounding notes. Migrating birds also utter a nasal note, though this is slightly different in character from that issued when perched. The flight, although similar to that of a Chaffinch, is more undulating and the white rump should be visible if you flush a bird from the ground.

winter male

winter male

breeding male

winter female

Migrating flock.

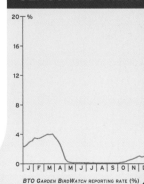

Winter female at a bird table.

WINTER ROOSTS

Huge numbers of wintering Bramblings may congregate together, with roosts numbering many millions of individuals recorded fairly often from parts of central Europe. One central European roost was estimated to number 20 million birds. Those noted in Britain and Ireland tend to be much smaller, not least because our wintering population only tops a million birds in the best of years. Even so, a roost in Merseyside exceeded 150,000 birds in one particular winter. Sadly many birds from this roost were killed by motor traffic, having first been incapacitated by the salt applied to the icy road surface from which they were taking fallen Beech mast.

A WINTER VISITOR

Bramblings leave their breeding grounds in September and move south, gathering in large flocks in areas with a plentiful supply of Beech mast and conifer seeds – favoured foods during the autumn and winter. Flocks often remain in these areas either until the supply of Beech mast is depleted or is covered with snow, at which stage they move further south. These movements bring Bramblings into Britain and (to a lesser extent) Ireland from mid-October. Unlike most other finches, Bramblings migrate at night and their arrival right along the east coast of Britain suggests that many make a direct crossing of the North Sea from Scandinavia. Because the availability of Beech mast can vary dramatically between years and areas, Brambling movements, and the numbers wintering in Britain, can be equally variable. In some winters we may have just 50,000 Bramblings present, in others in excess of 2 million may winter here. The Garden BirdWatch reporting rate reflects this and also shows the seasonal pattern to the use of gardens. When birds first arrive they are often unobtrusive, feeding on the ground in woodland but, as Beech mast becomes harder to find, they begin to exploit the food provided at garden feeding stations. Reporting of Bramblings in gardens begins in October but does not peak until late winter or early spring. Like Chaffinches, male Bramblings tend to winter farther north than females, so you may find that your wintering flock may be mainly of birds of one sex.

IT'S A FACT...

It is possible to build up the numbers visiting your garden by regular ground feeding with a mix of premium seed and peanut granules. This mix is also favoured by Chaffinches, with which Bramblings often associate, providing an ideal opportunity to practise your identification skills.

STATUS

Green listed because of favourable European situation.

SEASONAL TRENDS

20 %
16
12
8
4
0

J F M A M J J A S O N D

BTO Garden BirdWatch REPORTING RATE (%) THROUGHOUT THE MONTHS OF THE YEAR

GREENFINCH
Carduelis chloris

FAMILY
Carduelidae

LENGTH
15cm

WINGSPAN
26cm

WEIGHT
28g

HABITAT
Farmland, open woodland, gardens and parks.

FOOD
Large seeds, including rosehips. Nestling diet includes some invertebrates.

NEST
A robust base of twigs and grasses, lined with finer grass, roots, hair and feathers, usually in the fork of a bush or tree, early season preference for conifers.

EGGS
4-5 eggs, grey or bluish white, with sparse red, purple or black spotting.

EGG SIZE
20.6 × 14.8mm

INCUBATION
14-15 days. 2 broods.

BREEDING SEASON
April to September

POPULATION SIZE
695,000 breeding territories.

TYPICAL LIFESPAN
2 years

MAX. RECORDED AGE
12 years

The Greenfinch is a species that has learnt to utilise gardens and the food provided at feeding stations. Historically, Greenfinches were largely confined to areas of woodland or forest edge and were rarely seen in farmland or around human habitation. In some parts of continental Europe this is still the case but here, in Britain and Ireland, Greenfinches started visiting gardens in the early 1900s and are now one of the most familiar garden birds.

male

female

DISEASE

A disease, called Trichomonosis, previously associated with doves and pigeons, was first diagnosed in Greenfinches in 2005 and has since gone on to hit populations in various parts of the country, typically peaking in late summer. The disease is caused by a parasite, which is unable to survive for long periods outside of the host. This means that transmission of the parasite between birds is most likely to occur through regurgitated food or through contamination of drinking water with saliva from an infected bird. Infected individuals appear lethargic, with fluffed-up plumage. They may also drool saliva, have difficulty in swallowing and may have a severe swelling of the neck.

SHOULD I STAY OR SHOULD I GO?

Most of the Greenfinches that breed in Britain and Ireland are sedentary in their lifestyle, remaining close to where they were born. However, a small number move longer distances, leaving the breeding areas in autumn to winter in Ireland or departing from southeast England to winter on the Continent. These movements have aroused a great deal of interest among researchers because an individual bird may behave differently in different years. This fact, together with the timing of movements, suggests that the movements are made in response to high breeding densities or food shortages during late summer, rather than a response to weather conditions. At the same time as these Greenfinches are departing from Britain, others (typically only a small number) are arriving from the Continent, mostly from Norway. Some of these winter visitors are believed to be birds that normally would have wintered on the Norwegian coast, while others, arriving mainly in eastern England, are birds that have drifted too far west from their usual migration route, southwards down the coast of mainland Europe. Again, at the individual level, such movements appear to be occasional rather than annual, illustrating the complexities of bird migration and the great many questions that still remain to be answered by ornithologists. Long term studies undertaken by BTO researchers and volunteers will be crucial in helping to achieve this goal.

IT'S A FACT...

Greenfinch nests, typically placed between 1.5 and 5m off the ground, are quite bulky and are relatively easy to find in all but the densest bushes.

Greenfinches fighting.

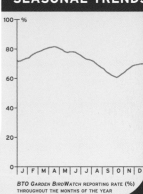

adult male

Flying birds show striking wing and tail patterns.

FOODS AND FEEDING

Like other members of the finch family, Greenfinches eat a lot of seeds and their large robust bills enable them to tackle a very wide range. However, Greenfinches show some preference for seeds held within a fleshy fruit and they adore rosehips, though they may disregard the flesh and just eat the seeds. In woodland, in addition to rosehips, Greenfinches also take other large seeds: from elm and Dog's Mercury in summer, Yew and Hawthorn in autumn and Bramble in winter. In farmland they take other seeds, favouring members of the crucifer and daisy families, many of which, such as Charlock, are less plentiful than they used to be. In gardens, Greenfinches have exploited sunflower seeds in a big way, favouring black sunflower seed and sunflower hearts over other foods on offer. Individual birds will sit at feeders and eat a steady succession of black sunflower seeds, expertly splitting them open with their sharp-edged bills and rejecting those too small to bother with.

BREEDING ECOLOGY

Although male Greenfinches may begin to sing as early in the year as January, they do not usually pair until late February or early March, with most clutches laid towards the end of April. Those birds nesting in gardens where food is provided throughout the year appear to be able to begin breeding slightly earlier than pairs elsewhere. This also means that they may be able to fit more broods into the breeding season and rear more young. Greenfinches often nest in loose colonies of four to six nests and the birds do not really defend clear territories in the way that most other species do. Instead, they only defend the area immediately around the nest, often foraging some distance away. This means that they will often feature at bird feeders in gardens that appear to have no suitable nesting opportunities close by. The growing young are fed on regurgitated food, made up of seeds and some insects, so you are unlikely to see adults carrying food in their bills.

juvenile

IT'S A FACT...

The planting of ornamental conifers in parks and gardens has provided Greenfinches with excellent nesting opportunities.

STATUS

Green listed.

SEASONAL TRENDS

BTO GARDEN BIRDWATCH REPORTING RATE (%) THROUGHOUT THE MONTHS OF THE YEAR

BTO
BIRDFACTS

FAMILY
Carduelidae

LENGTH
12cm

WINGSPAN
24cm

WEIGHT
17g

HABITAT
Woodland, farmland and open country.

FOOD
Small seeds, with some invertebrates taken in the summer.

NEST
A very neat nest composed of fine grasses, moss, roots and spider silk, well-hidden in branches of bush or tree.

EGGS
4-6 eggs, pale bluish-white in colour and scrawled with fine red markings.

EGG SIZE
18.3 × 13.4mm

INCUBATION
13-15 days. 2, sometimes 3, broods.

BREEDING SEASON
April to September

POPULATION SIZE
299,000 breeding territories.

TYPICAL LIFESPAN
2 years

MAX. RECORDED AGE
8 years 8 months

GOLDFINCH
Carduelis carduelis

The Goldfinch is a small, delicate and beautifully marked finch, its colourful plumage bright and characteristic, and its agile feeding behaviour a joy to watch.

THE IMPORTANCE OF PLANTS

Goldfinches are widespread within Britain and Ireland but are most abundant in lowland areas, where favoured food plants are most numerous. Many of the plant species used belong to the family Compositae and include familiar species such as dandelions, groundsels and ragworts. It is the seeds of these plants upon which the Goldfinches feed, preferring to take those that are not fully ripe but still in a milky state. This preference means that Goldfinches are quite mobile feeders, moving to new areas to find plants with seeds in a suitable state. The seasonal pattern to the Garden BirdWatch reporting rate for this species is probably influenced by the availability of different seeds. Historically, Goldfinches appear to have made use of natural foods within gardens at those times of year (especially late spring) when seed supplies in other habitats were low. In more recent years, Goldfinches have made increasing use of garden feeding stations, exploiting foods like sunflower hearts and niger seed. Alongside this change in feeding behaviour, we have seen an increase in the peak Garden BirdWatch reporting from 23% to 60% of gardens over an 12-year period. Goldfinches have bills that are quite long and thin, enabling them to extract seeds from plants that are typically not available to other finches.

The Goldfinch is the only finch to be able to extract seeds from teasels, although even female Goldfinches find this difficult because their bills are slightly shorter than those of the males.

Adults are colourful and striking in flight.

Adult feeding on thistle seeds.

SPRING FEEDING

Traditionally, garden bird feeding has stopped once the last snow and ice has gone. However, April is a really important month for the Goldfinch and other seed-eaters. By continuing to feed through the spring, you will be helping these endearing little birds at a time of the year when their other food sources are hard to find.

Adult on Teasel seedhead.

IT'S A FACT...

Young Goldfinches lack the black, red and white facial markings of the adults but do show the black wings with prominent yellow bar.

juvenile

Adult landing.

Adult feeding on niger seeds.

Perched adult.

MIGRATION

Although Goldfinches may be seen in gardens throughout the year, many leave Britain in the autumn to winter on the Continent. These migrants do not appear to move to a specific wintering area, instead they seem to migrate south and stop once they find suitable conditions. Many Goldfinches remain within Britain and Ireland during the winter, some close to their breeding grounds while others move south or cross into Ireland. Researchers examining the sex ratios of Goldfinches remaining within Britain during the winter have produced evidence that suggests more females migrate than males and that the females winter further south. This is a pattern also seen in some other finches, for example Chaffinch.

BEHIND BARS

So endearing are these delightful little birds that they have proved to be very popular as cagebirds, a fashion which put great strain on the population throughout the 19th century. Although the live-trapping of Goldfinches has been banned for many decades in Britain and Ireland, the practice still goes on in some parts of Europe, where finches are caught in automatic traps known as Chardonnerets – 'Chardonneret' also being the French word for Goldfinch.

NESTING BEHAVIOUR

Goldfinches often site their nests higher off the ground than other finches, sometimes placing them in the fork of a branch up to 15m high. The nest itself is similar to that of the Chaffinch, neat and compact with a deep cup. The egg-laying period extends from late April through to August, with young in the nest as late as September, giving many pairs the opportunity to rear two or even three broods if conditions are favourable. The developing nestlings are fed a mixture of regurgitated seeds and insects, with early broods receiving more insects than late ones, a reflection of both insect and seed abundance. Some gardens may also hold nesting pairs, which defend only a very small territory around the nest site and usually feed some distance away. In urban and suburban areas, the presence of waste ground (with its many weeds) may support local breeding pairs and the increasing clean-up of such 'brownfield' sites may be reducing opportunities for Goldfinches. However, the specialisation of Goldfinches on particular plant species and their seeds may be one reason why the Goldfinch population in Britain and Ireland has not declined in the manner that other species feeding on different plant species have done.

STATUS

Currently increasing, hence Green listed.

SEASONAL TRENDS

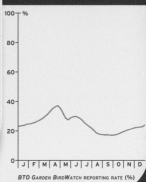

BTO GARDEN BIRDWATCH REPORTING RATE (%) THROUGHOUT THE MONTHS OF THE YEAR

BTO BIRDFACTS

SISKIN
Carduelis spinus

Male feeding on bird feeder.

Back in the early 1970s, the Siskin was an unfamiliar species to many birdwatchers, restricted as a breeding species to Scotland and to those areas with large conifer plantations within England and Wales. However, it is now a familiar winter visitor to many garden bird tables away from the expanding breeding range.

THE MOVE INTO GARDENS

Breeding Siskins are associated with conifers and used to be largely restricted to Caledonian pine forests in northern Britain. However, the widespread establishment of plantation forest (notably Sitka Spruce) has greatly increased the amount of suitable breeding habitat. This increase in the breeding population has been matched by a change in feeding behaviour during winter. Siskins have been recorded visiting gardens since the mid-1960s, initially feeding on fat but later learning to exploit peanuts and sunflower hearts. Prior to developing the habit of exploiting garden feeding stations, Siskins mainly wintered in alder, birch and larch woodlands, often associating with Redpolls. It has been suggested that Siskins first moved into gardens to exploit the seeds of introduced exotic conifers, established by gardeners to provide winter greenery, and that this then led to them discovering the supplementary foods put out by Garden Birdwatchers. As can be seen from the annual Garden BirdWatch reporting rate, garden feeding is a feature of late winter and early spring, suggesting that the birds move into gardens once natural seed crops have been depleted. Garden feeding may also provide an advantage during early mornings or on wet or overcast days, when cones are closed and the seeds unobtainable. It is only during dry conditions that the cones open up to reveal the seeds within.

Male feeding on Common Nettle seeds.

FAMILY	Carduelidae
LENGTH	12cm
WINGSPAN	22cm
WEIGHT	15g
HABITAT	Coniferous woodland, gardens in winter.
FOOD	Seeds, especially those of spruce, pine, alder and birch.
NEST	Delicate structure, made from twigs, heather, rootlets and grass; placed high in conifers among the outer branches.
EGGS	4-5 eggs, bluish-white or blue and marked with brown spots and scrawls.
EGG SIZE	16 × 12mm
INCUBATION	12-13 days. 2 broods.
BREEDING SEASON	April to August
POPULATION SIZE	357,000 breeding territories.
TYPICAL LIFESPAN	Not known
MAX. RECORDED AGE	9 years 2 months

female

IDENTIFICATION

The Siskin is smaller than a Goldfinch and much smaller than a Greenfinch. The plumage is predominantly a yellow-green colour, with a striking yellow band on the wing and yellow patches at the base of the tail. Adult males have a black crown and a lot of black in the wing. Females are greyer in colour, streaked above and without the black crown. Juveniles resemble females but are buff-brown above and more heavily streaked (both above and below).

IT'S A FACT...

Nests are usually placed high in conifers among the outer branches, making them virtually inaccessible. They are delicate in structure and hemispherical in shape, made from twigs, heather, grass and spiders' webs, woven together and lined with rootlets and hair.

BOOM AND BUST

Production of seed by coniferous trees like the spruces and pines is known to vary from one year to the next. Being so dependent upon these seeds means that both Siskin breeding success and movements are determined by seed availability. In years following poor seed crops, Siskins start breeding much later in the year and rear fewer young. Variations in food availability also lead to eruptive movements of Siskins from breeding populations in the conifer forests of Scandinavia, bringing birds to Britain in varying numbers in different years. In most years, between 15 and 20% of gardens participating in the BTO Garden BirdWatch report visiting Siskins; however, in those years when conifer seed is in short supply this can leap to 40% of gardens.

WINTER MOVEMENTS

Although some Siskins (notably those in northern regions and in Ireland) winter close to the breeding areas, many others move south to winter in central and southern Britain. At the same time, birds from Scandinavia and elsewhere in continental Europe enter Britain, either across the English Channel via Belgium and the Netherlands, or by making a direct crossing of the North Sea. Some of these birds remain here for the winter, while others pass through to wintering grounds elsewhere in Europe and North Africa. Individual birds may return to the same wintering area in successive years or may winter in widely different locations from one year to the next. Recent research suggests that, in some populations, individuals may adopt one of two different approaches to wintering. Some individuals settle very quickly in an area and effectively become residents, only making short-distance local movements. Others, perhaps the vast majority in some populations, become transients and are very mobile, covering large distances. Siskins visiting suburban gardens in southern Britain often arrive in small parties and a succession of these groups will pass through individual gardens during the same day. Most Siskins will have left the gardens of southern Britain by mid-April, with birds returning surprisingly quickly to their breeding grounds. One Siskin, ringed at a site in Shropshire, was recovered in the Highland region of Scotland just three days later, thus averaging 190km per day.

Perched acrobatically on winter twigs.

IT'S A FACT...

The Siskin showed a dramatic expansion in its breeding and wintering distribution within Britain and Ireland over the course of the 20th century.

male

Male feeding on niger seeds.

STATUS

Green listed.

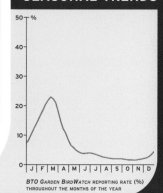

SEASONAL TRENDS

BTO GARDEN BIRDWATCH REPORTING RATE (%) THROUGHOUT THE MONTHS OF THE YEAR

BTO BIRDFACTS

FAMILY
Carduelidae

LENGTH
16cm

WINGSPAN
26cm

WEIGHT
21g

HABITAT
Woodland and wooded farmland.

FOOD
Seeds of fleshy fruits, together with buds and shoots. Young are fed on invertebrates.

NEST
Base made of small twigs, upon which a woven cup of rootlets is constructed. Placed in thick cover.

EGGS
4-5 eggs, pale blue with dark purple spots.

EGG SIZE
20.2 × 14.6mm

INCUBATION
14-16 days.
2, sometimes
3 broods.

BREEDING SEASON
April to September

POPULATION SIZE
158,000 breeding pairs.

TYPICAL LIFESPAN
2 years

MAX. RECORDED AGE
9 years 2 months

BULLFINCH
Pyrrhula pyrrhula

Despite their characteristic and striking appearance, Bullfinches are relatively shy birds. Breeding pairs occupy a range of habitats, characterised by the presence of dense bushes in which they nest, and are found across much of Britain and Ireland. They are absent as a breeding species from the Isle of Man, the Western Isles, Orkney and Shetland.

A WOODLAND BIRD

The Bullfinch is one of our most enigmatic birds, not least because of its shy and retiring nature. Although the Bullfinch favours deciduous woodland as its habitat of choice, since the late 1990s this species has made increasing use of garden feeding stations. Small parties are reported visiting gardens across much of Britain and Ireland, with rural (and increasingly suburban gardens) favoured over those of a more urban nature. This reflects a preference for cover, especially for nest site selection, and it is those gardens that are connected to small areas of woodland or scrub by thick hedgerows that seem the most attractive. Even so, the BTO Garden BirdWatch reporting rate is low when compared with other familiar garden birds. Bullfinches generally remain in the same area throughout their lives, although they may move away from breeding sites during the winter months in order to exploit feeding opportunities elsewhere.

IT'S A FACT...

Bullfinch pairs appear to be long-lasting, with individuals remaining together from one breeding season to the next. Established pairs do not seem to be overly territorial during the breeding season, perhaps because of the low density at which they occur.

male

female

IT'S A FACT...

For much of the year, Bullfinches feed on the seeds of Ash, elms, Common Nettle, docks and Dog's Mercury, among others, only taking insects when feeding young in the breeding season.

FEEDING TECHNIQUES

The Bullfinch appears slower and more deliberate than other finches when feeding. Individuals can be seen manipulating seeds and buds in the bill, turning food items with the tongue and removing the pulp against the lower mandible before swallowing the seeds. The birds prefer to feed on the plant, and often leave the skin and pulpy flesh of a fruit hanging, having removed the seeds. In recent years they have started to utilise seed feeders, taking black sunflowers and other seeds.

Female feeding on fallen seeds below a feeder.

male

male

female

Male in flight.

REARING A FAMILY

Very few pairs of Bullfinches breed within gardens, simply because the majority of gardens do not have sufficient thick cover within which the birds can nest. The nest itself is usually placed just inside the canopy of a suitable shrub or bush (such as Bramble, Hawthorn or Blackthorn), some 1-2m off the ground. Adult Bullfinches may feed at some distance from the nest, something that is forced upon them because of the patchy distribution of their favoured food. To help the parents bring back food to their chicks, they have special food sacs positioned in the floor of the mouth. Bullfinches are the only finches with such sacs.

BULLFINCHES AND BUDS

Between the 1950s and mid-1970s, Bullfinches were abundant across much of southern Britain, in some areas becoming a serious pest of commercial fruit trees. This led to licensed control of the birds, something that seems to have had no significant effects on population size. The more recent decline in numbers, highlighted by BTO survey work, is thought to be the result of changing agricultural practices and the loss of arable weeds. Culling of Bullfinches has now virtually ceased as the species is no longer the pest it once was. Bullfinches turn to the buds of fruiting trees in late winter and early spring, when supplies of tree seeds, especially Ash, run low. Flower buds are preferred to leaf buds because they are nutritionally more rewarding. Selection of cultivated varieties over uncultivated varieties occurs because Man has selectively bred the cultivated varieties to give a higher yield, making them more appealing to the Bullfinches.

STATUS

Red listed because of long-term decline.

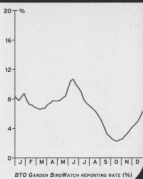

SEASONAL TRENDS

BTO GARDEN BIRDWATCH REPORTING RATE (%) THROUGHOUT THE MONTHS OF THE YEAR

LESSER REDPOLL
Carduelis cabaret

FAMILY	Carduelidae
LENGTH	12cm
WINGSPAN	22cm
WEIGHT	11g
HABITAT	Woodland, birch/willow scrub.
FOOD	Very small seeds, especially birch. Invertebrates taken in summer.
NEST	A foundation of twigs on which a densely packed inner layer of grass and roots sits. This is lined with hair, wool and feathers. Sited in shrub or tree.
EGGS	4-5 eggs; bluish-white in colour with variable rust-red blotches and brown scrawls.
EGG SIZE	16.9 × 12.6mm
INCUBATION	11-12 days. 2 broods
BREEDING SEASON	May to July
POPULATION SIZE	25,000 pairs.
TYPICAL LIFESPAN	Insufficient data
MAX. RECORDED AGE	6 years

The Redpoll is a very agile feeder that specialises in taking the small seeds of birch and, to a lesser extent, Alder. Primarily a bird of birch woodland and young conifer plantations, the Lesser Redpoll (to give it its proper name) may visit gardens in winter, feeding alongside Siskins on mixes containing suitably small seeds.

male

female

female

WHAT'S IN A NAME?

The Lesser Redpoll was, until recently, considered to be a subspecies of the Common Redpoll (*see* box). However, differences in plumage, behaviour and voice have long suggested that the Lesser Redpoll was really a separate species. This was confirmed in the early 1990s, when an expanding breeding population brought these delightful little birds into contact with Common Redpolls breeding in southern Scandinavia. Researchers found that the two species bred alongside each other and did not interbreed. As a result of this work, and a review of the other differences that existed between the two birds, a decision was made in 2001 to treat them as separate species. The Lesser Redpoll breeds across Britain and Ireland and in those countries bordering the North Sea, from France north to southern Norway. The larger Common Redpoll breeds in the birch forests of Northern Europe, with small numbers arriving in Britain as winter visitors. So, despite the English names, the Common Redpoll is not common here but the smaller Lesser Redpoll is!

ON THE MOVE

During the autumn and winter there is a noticeable retreat from higher ground, with many of our breeding Lesser Redpolls moving south to winter in southern Britain. Some continue further south, wintering on the near Continent. The extent of these movements appears to be dependent upon food availability and competition for the available resources. If the autumn birch seed crop is poor and there is also a sizeable post-breeding population, then many more birds will move south. There are years when huge numbers of birds leave Britain altogether, with some ringed individuals reaching as far away as northern Italy, southern France and Portugal.

male

female

RAISING A FAMILY

Although the species nests primarily in scrub and birch woodland, it may sometimes breed in larger gardens. Interestingly, Lesser Redpolls are semi-colonial breeders, which means that five or six pairs will nest in close proximity. Males advertise to prospective mates through a special song flight and it is worth listening out for the characteristic rattling call. Most nests are built in broad-leaved trees, though those built early in the season are almost invariably in conifers, which afford greater cover when many broad-leafed trees are still in bud. The young are fed on a diet of small seeds and insects, the latter being particularly important during the first week of life. Most of the food is taken from the tree canopy, the birds only really foraging on the ground later in the year when the seed crop has fallen to the ground. Once the first breeding attempt has finished, another will usually be made, although pairs may move location between the two attempts. At the end of the breeding season the Lesser Redpolls quickly form post-breeding flocks and initiate their annual moult, preparing for the winter ahead.

COMMON REDPOLL

Although the Common Redpoll *Carduelis flammea* breeds in the birch forests of Russia and northern Europe, it remains a regular visitor to Britain, primarily arriving on the east coast during October. The numbers present vary greatly from one year to the next, reflecting the availability of birch seed elsewhere in Europe. If the seed crop is very poor then unusually large numbers may arrive here. These winter visitors are noticeably larger and paler than the Lesser Redpolls that are largely resident in Britain and Ireland throughout the year.

Common Redpoll

male

IT'S A FACT...

Although the shape of the Lesser Redpoll's bill is similar to that of a Greenfinch it is noticeably smaller and has a very narrow tip, the latter being an adaptation to extracting birch seed.

female

UPS AND DOWNS

Because the Lesser Redpoll is a bird of pioneer woodlands, it favours areas of scrubby birch woods and the early stages of conifer plantations, where birch often grows alongside the newly-established conifers. This association has had a bearing on the size of the breeding population, with numbers increasing rapidly from the 1950s through to the 1970s as new areas of conifer plantation became established. With a slow-down in the creation of new plantations, coupled with the loss of birch as existing plantations matured, Redpoll populations have since declined.

IT'S A FACT...

Before the all-important birch seeds ripen in July, Lesser Redpolls feed predominantly on the flowers and seeds of Sallow, together with bud-dwelling insects and spiders.

STATUS

Has been moved onto Amber list because of population decline.

SEASONAL TRENDS

BTO GARDEN BIRDWATCH REPORTING RATE (%) THROUGHOUT THE MONTHS OF THE YEAR

BIRDFACTS

REED BUNTING
Emberiza schoeniclus

winter male

female

The Reed Bunting is a scarce visitor to gardens, only tending to occur in rural gardens during winter or late spring. In this respect it is similar to the Yellowhammer, our other familiar bunting species.

HABITAT PREFERENCES

Reed Buntings breed in a wide variety of habitats, not just the damp one suggested by the name. In fact, Reed Buntings can be found breeding among the marram grass of sand dune systems, in farm hedgerows and even within fields planted with oilseed rape. The use of drier habitats is a relatively recent phenomenon, possibly a consequence of the loss of damp habitats to agricultural intensification or because of the decline in Yellowhammer abundance (a potential competitor within drier habitats). Reed Bunting populations in lowland Britain have shown a similar pattern of decline to many other small seed-eating birds. Although Reed Buntings are almost exclusively insectivorous during the breeding season, they quickly revert to feeding on seeds in late summer. It seems likely that their decline is in part due to a lack of available seeds during the winter months. Evidence in support of this hypothesis comes from the Garden Bird Feeding Survey, which revealed a major increase in the use made of gardens by Reed Buntings during the period of decline for farmland populations. Although it is known that severe winter weather can have a significant impact on Reed Bunting populations, this effect is short-lived and recovery from weather-related declines happens quickly.

FAMILY	
Emberizidae	
LENGTH	
13cm	
WINGSPAN	
24cm	
WEIGHT	
21g	
HABITAT	
Farmland, damp habitats and scrub.	
FOOD	
Seeds and, in the breeding season, invertebrates.	
NEST	
Sedges, grasses and other waterside plants, lined with finer material. Placed low in cover.	
EGGS	
4-5 eggs, pale purple or olive-brown and marked with darker tones.	
EGG SIZE	
19.9 × 14.7mm	
INCUBATION	
13-14 days. 1-2 broods.	
BREEDING SEASON	
May to September	
POPULATION SIZE	
180,000 breeding territories.	
TYPICAL LIFESPAN	
3 years	
MAX. RECORDED AGE	
9 years 11 months	

IDENTIFICATION

Reed Buntings may be missed in winter. The characteristic black head and white collar of the breeding male are hidden behind brown feather tips that will gradually be worn away in spring to reveal the striking breeding plumage. Females are less boldly marked than the males.

IT'S A FACT...

It is during the winter months that large flocks of Reed Buntings often gather together to roost, using wet or marshy areas and reedbeds, where they will be able to spend the night away from likely predators. During the day, these roosts break up and the birds forage widely over farmland and waste ground, in search of weed seeds and spilt cereal grain.

Breeding plumage male feeding on sunflower seeds at a bird table.

breeding plumage male

BREEDING BEHAVIOUR

Reed Buntings have a long breeding season and eggs can be laid from early May, typically in a well-concealed nest positioned low down in the vegetation. Adult males usually establish their breeding territory on the same site as used the previous year, a pattern repeated in adult females though not as strongly. Unlike the Yellowhammer, the Reed Bunting does not need tall shrubs and trees from which to proclaim its territorial rights. Instead, it will readily sing from the stem of a reed or dock just protruding above the surrounding vegetation.

IT'S A FACT...

Reed Buntings surprised near the nest may feign injury in an attempt to draw you away from where the nest is hidden.

winter male

female

THE VALUE OF OILSEED RAPE

Reed Bunting densities are four times higher in oilseed rape fields than they are in cereals or set-aside, highlighting the value of this crop for the species. These fields of oilseed rape can offer good foraging and nesting opportunities early in the season but by early summer they are either cut or sprayed with herbicide. Spraying kills the plants and allows them to dry prior to harvest. Second broods of Reed Buntings are killed if the field is cut but they appear to be able to survive herbicide spraying, giving the birds a window of an additional two weeks to get their young fledged.

STATUS

Despite recent increase, still Red listed because of long-term decline.

breeding plumage male

WINTER MOVEMENTS

The vast majority of British and Irish Reed Buntings remain here in the winter and are joined by a very small number of birds from Scandinavia. Winter can be a difficult time for these birds, especially when the seeds they feed upon become unavailable because of frost or ice. At such times they may turn to garden feeding stations, though not venturing far and so are only likely to turn up at rural gardens. Even hungry Reed Buntings rarely take to hanging feeders, preferring instead to take seed scattered on the ground. When conditions improve they will return to those habitats they were foraging within before the bad weather. Further north, the extent to which populations of Reed Buntings migrate seems to be linked to temperature and the degree of snow cover. Where the mean January temperature is below 0°C, the birds migrate; where it is above 5°C the population is essentially sedentary.

SEASONAL TRENDS

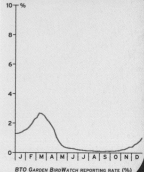

BTO GARDEN BIRDWATCH REPORTING RATE (%) THROUGHOUT THE MONTHS OF THE YEAR

JUST PASSING THROUGH

Although not regular garden visitors, a number of migrant birds – summer visitors to Britain – make use of gardens on occasion. In spring, adults of several species may turn up in a range of habitats not necessarily used for breeding. In autumn, both adults and young of the year may visit gardens to feed on invertebrates or fruits and berries. The sugars in fruits and berries provide an accessible energy source, helping the birds to build up reserves prior to migration.

NIGHTINGALE

Luscinia megarhynchos

Nightingale

Famed for its rich and largely nocturnal song, the Nightingale is a summer visitor to the southeast of England. Here it breeds in areas of scrubby woodland and more open scrub, with dense bramble. While a few garden birdwatchers may be fortunate enough to have breeding Nightingales nesting close to their rural garden, the species remains a rare garden visitor. Nightingales are not the only birds to sing at night; any bird that you hear singing in the small hours from your urban or suburban garden will almost certainly not be a Nightingale. It is likely to be a Blackbird, Song Thrush or Robin.

REDSTART

Phoenicurus phoenicurus

The Redstart is one of the characteristic birds of the western oak woods that stretch from Cornwall, up through Wales and across northern Britain. A summer migrant, arriving in April, the Redstart nests in holes in mature trees. As such, it takes readily to nestboxes, provided they are placed in wooded habitats with sufficient numbers of insects (especially moths and their caterpillars) available to them. Gardens located alongside suitable breeding habitat may be used throughout the breeding season but those elsewhere are only likely to be visited briefly by birds passing through on migration.

BELOW: **Redstart male**

Redstart male

Nightingale

Redstart female

Reed Warbler

Reed Warbler

REED WARBLER

Acrocephalus scirpaceus

As the name suggests, this species is predominantly found breeding in reedbeds. However, in recent years it has been increasingly recorded breeding in fields of oilseed rape and well-developed stands of willowherb species. The birds arrive back from their African wintering grounds from mid-April and soon establish breeding territories, whose ownership is advertised through a chattering song. Reed Warblers build a nest that is woven around the vertical stalks of the vegetation within which it is placed, often over water to reduce access to predators. Despite these efforts, the Reed Warbler is one of the main species targeted by Cuckoo.

SEDGE WARBLER

Acrocephalus schoenobaenus

Like its relative, the Reed Warbler, this species is a summer visitor most commonly associated with wetland reedbeds. However, the Sedge Warbler will also breed in dry habitats, such as Bramble or Hawthorn thickets, occasionally using oilseed rape crops and young conifers. The species has a wider breeding distribution across Britain, reaching northern Scotland and Orkney (Reed Warbler only just makes it into Scotland), and may, on occasion, turn up in a garden. The song is superficially similar to that of the Reed Warbler, but breaks from the regular chatty tempo into more explosive phrases as if the bird has suddenly become over-excited.

Sedge Warbler

WILLOW WARBLER

Phylloscopus trochilus

The Willow Warbler is perhaps our most common summer visitor, breeding right across Britain and Ireland within a range of woodland and scrubby habitats. Gardens are only rarely used for breeding and those that are tend to be rural in nature and rather overgrown. Similar in appearance to the Chiffchaff (*see* page 160), the Willow Warbler is best separated by its song, a cascading series of notes that is very different from the disyllabic 'chiff-chaff' of its relative. Thanks to the efforts of bird ringers we know that our Willow Warblers migrate south through France and Spain, crossing the Sahara to winter in the forests of West Africa.

LESSER WHITETHROAT

Sylvia curruca

This is the least conspicuous of our visiting *Sylvia* warblers (Blackcap, Whitethroat and Garden Warbler) and, with its tendency to skulk and deliver its song from cover, it is often overlooked. Mature hedgerows, patches of scrub and young plantations are favoured within its British breeding range. However, elsewhere in Europe the species seems to be increasingly associated with human sites, utilising the scrubby habitats available in larger gardens, parks and cemeteries. Unusually among our summer visitors, the Lesser Whitethroat is one of the few species to migrate southeast in autumn, wintering at the eastern end of the Mediterranean. Most of our other visitors migrate southwest, taking them to, or through, the western end of the Mediterranean.

Lesser Whitethroat

Willow Warbler adult

Willow Warbler juvenile

REGIONAL SPECIALITIES

A number of bird species have restricted ranges in Britain, either because they are associated with particular habitats, have very small breeding populations or are constrained by other factors. Some of these species may visit gardens within the area that they are found, using the food we provide (for example, Cirl Bunting) or the nesting opportunities on offer (for example, Herring Gull).

HERRING GULL *Larus argentatus*

While we tend to think of gulls as seabirds, the Herring Gull is one of two species to have established sizeable breeding colonies at inland sites (the other is Lesser Black-backed Gull, which is the more common of the two species). Roof-nesting by Herring Gulls was first noted in 1910 but has since increased, with an estimated 12,000 pairs roof-nesting in England and a further 6,000 pairs in Scotland. The largest roof-nesting colonies are to be found in Aberdeen, Bristol and Cardiff. It is birds from these urban colonies that are most likely to visit gardens to take scraps from bird tables.

Herring Gull

RED KITE *Milvus milvus*

Although once a common sight across much of Britain, by 1900 the Red Kite population had been reduced by persecution to a tiny relict population inhabiting the hanging oakwoods of mid-Wales. Thanks to the efforts of conservationists, the species has since been successfully reintroduced to a number of sites across the country. Birds from some of these breeding populations are now regularly sighted at garden feeding stations in some areas (for example, in parts of the Chilterns), where they take scraps put out by delighted garden birdwatchers. Kites are scavengers and take a wide range of food items, from small birds and rodents, through to various invertebrates and carrion, the latter being particularly important during winter. Their large nests are built in woodland and scavenged items, including rags and other soft furnishings, may be incorporated into the nest.

Herring Gull

Red Kite

PIED FLYCATCHER

Ficedula hypoleuca

RIGHT: **Pied Flycatcher male**

With its black and white plumage, the male Pied Flycatcher is a striking summer visitor to oak woodlands across the western half of Britain. Arriving sometime between mid-April and late May, this cavity-nesting species is a rare visitor to gardens. Rural gardens located alongside suitable breeding habitat may be used, with gardens outside of the core breeding range only being visited during the period of spring or autumn migration. Being a summer visitor means that each female is only able to rear a single brood before having to depart again for her African wintering grounds. Males, on the other hand, are polygamous and may have two or even three females on the go. The male diverts most of his energies to his initial mate, the others being left to rear the brood largely on their own.

CRESTED TIT *Lophophanes cristatus*

The Crested Tit has a greatly restricted distribution, occurring only within a small part of the Scottish Highlands, where it favours native pinewoods and mature plantations. The adults are extremely sedentary in their habits and typically remain within their breeding territories throughout the year. Young birds, dispersing away from where they were born, may result in individuals appearing in atypical habitats, including gardens, and there have been recent records of such individuals visiting garden feeding stations. The use of garden feeding stations may spread, as has been seen in other species, but the degree to which this might happen is impossible to predict. The tiny breeding population is thought to number between 1,000 and 2,000 pairs.

Pied Flycatcher female

CIRL BUNTING *Emberiza cirlus*

Although never common, the Cirl Bunting is now one of our rarest breeding songbirds, occupying a small part of its former breeding range (which once extended across the extreme south of England). The species probably first reached us during the 18th century but the second half of the 20th century saw a massive decline in numbers, with the population falling to just over 100 breeding pairs. Since then, a major conservation effort has been directed towards the species, with the provision of supplementary food and appropriate habitat management helping to increase breeding numbers. Within the still small breeding range, some garden feeding stations located in rural areas have been well used during winter. Ironically, the future of the species is still uncertain, with the threat of urban expansion likely to result in loss of the mixed farmland habitats needed by the Cirl Bunting.

Cirl Bunting male

ABOVE: **Crested Tit**

Crested Tit

Cirl Bunting male

MALLARD *Anas platyrhynchos*

Instantly recognisable, the Mallard is the most widely distributed of all our ducks. This reflects an adaptable and opportunistic nature, with Mallards able to breed on a very wide range of waterbodies; they can also breed some distance from water, even in garden flowerbeds. Pairs may suddenly arrive in gardens during spring, remaining for several weeks before disappearing to breed nearby. Our breeding Mallards are largely sedentary in nature, with many thousands bred in captivity and then released for shooting each year. During winter, these birds are joined by arrivals from populations breeding in other European countries. Many Mallards appear indifferent to people and will take food from the hand. Urban populations are thought to rely on the food provided by Man.

GREY HERON
Ardea cinerea

The Grey Heron is an occasional and largely unwelcome visitor to garden ponds, where it may rapidly decimate any stock of goldfish or koi carp. Such fish tend to make easy pickings because garden ponds are usually fairly shallow and lack the vegetative cover within which the fish can hide. It surprises many to learn that this long-legged hunter builds large stick nests in the tops of mature trees. The nests are usually found in small colonies (known as heronries), which tend to be somewhat noisy affairs. The changing fortunes of our Grey Heron population has been monitored through the BTO's annual Heronries Census, first started in 1928 and one of the longest running bird surveys in the world.

juvenile

adult

juvenile

adult

TOP AND ABOVE: **male**

male

male

female

BUZZARD *Buteo buteo*

The Buzzard has shown a remarkable resurgence in recent years. Heavily persecuted, breeding Buzzards were once absent from large parts of the country, only hanging on in the extreme north and west. Now that this persecution has decreased markedly, they are becoming an increasingly common sight over large areas, nesting in woodland and foraging over open habitats. Although Buzzards may be seen circling overhead, they rarely venture into gardens, only doing so if weather conditions are particularly poor. Even then, it only tends to be large open gardens in rural areas that are visited.

KESTREL *Falco tinnunculus*

The sight of a Kestrel hovering into the wind above a roadside verge is familiar to many but this is a bird of prey that only rarely visits gardens. Unlike the Sparrowhawk that catches songbirds and hence may be attracted to garden feeders, the Kestrel takes mostly small mammals from areas of rough grassland (hardly typical garden habitat). As with the Buzzard, larger rural gardens are favoured, more so if they happen to have a paddock of rough grassland alongside them. Kestrels may use a large open-fronted nestbox, positioned in a hedgerow tree near suitable hunting habitat.

ABOVE: **female**   RIGHT: **male**

WATER RAIL *Rallus aquaticus*

Although widely distributed, this denizen of dense marshy vegetation is one of our most elusive birds. As such, it is more often heard than seen, with its characteristic pig-like grunts and squeals, delivered at dawn or dusk. Water Rails are omnivorous and although they feed mainly on aquatic insects, they will also take amphibians, plant material and carrion. During spells of very cold weather their favoured foraging habitats may freeze over, forcing individuals to seek food elsewhere. This is when Water Rails are most often reported from gardens, dashing out from cover to feed on kitchen scraps alongside some of the more familiar garden birds. Water Rails may sometimes kill and eat the smaller songbirds that feed alongside them.

MOORHEN *Gallinula chloropus*

The Moorhen is a species that is most commonly associated with still or slow-moving waterbodies within lowland landscapes. Here, a wide range of waterbodies may be used for breeding, including larger garden ponds. Breeding takes place between the middle of March and the end of August, with a third or more of breeding pairs rearing more than one brood of chicks. Sometimes, the young from the first brood will remain with their parents to help rear the second brood, a behaviour that provides them with experience and improves the prospects of their siblings. Moorhens are recorded regularly at a small proportion of garden feeding stations, where they feed on scattered bread and grain.

adult

Adult with chicks.

adult   **juvenile**

STOCK DOVE

Columba oenas

To some extent, this medium-sized, grey and black, dove resembles the larger Woodpigeon with which it sometimes associates; for this reason, it is often overlooked. While the expanding Woodpigeon population has brought it into gardens in increasing numbers, the Stock Dove remains a scarce visitor, largely restricted to more rural gardens. Stock Doves nest in cavities, both in trees and walls, as well as in old Rabbit burrows and the old nests of other species. Parkland and farmland with old trees are the preferred breeding and feeding habitats but even here the birds may be overlooked.

Juvenile in Reed Warbler nest.

CUCKOO

Cuculus canorus

Renowned for its habit of laying its eggs in the nests of other species, and familiar for its unmistakable song, the Cuckoo is one of our best known summer visitors. Males arrive from mid-April and may continue to call through until the end of June. Females arrive a few weeks later and return to the same general area in subsequent years. Individual females specialise in one particular host, the most commonly exploited being Dunnock, Meadow Pipit and Reed Warbler. This need for specialisation comes about because the eggs of the host species differ in colour and pattern and the Cuckoo needs to get a good match to avoid her egg being spotted by the owners of the nest into which it has been placed. Numbers have declined in recent years, possibly reflecting declining populations of the host species.

adult

Stock Dove

BELOW: **adults**

Turtle Dove

Turtle Dove

TURTLE DOVE *Streptopelia turtur*

The gentle purring call, which can sound almost frog-like when distant, heralds the spring arrival of our only migrant dove. Preferring warm and dry conditions, the Turtle Dove is largely restricted to Eastern England, favouring areas of arable land and well-developed hedgerows. Numbers have declined dramatically since the 1970s and the species is now scarce or absent in many of its former haunts. A few favoured garden feeding stations are visited by Turtle Doves, the birds feeding on seed mixes, either presented on the ground or on low bird tables. Despite legal protection within Europe, many thousands of Turtle Doves are shot each year on migration (both in southern Europe and in North Africa), a factor that may place additional pressure on this declining species.

Turtle Dove

BARN OWL *Tyto alba*

The Barn Owl is a lowland species, favouring areas of rough and unimproved grassland over which it can hunt for shrews, mice and voles. Barn Owls nest in large tree cavities and readily exploit large nestboxes erected for their use. The loss of suitable nesting and foraging sites resulted in a decline in the breeding population last century, but more recent conservation efforts have helped to stabilise the population. Barn Owl boxes or lofts, erected when barns are converted into dwellings, may enable breeding pairs to remain in the local area and such pairs often breed successfully alongside their new human neighbours.

Barn Owl

LESSER SPOTTED WOODPECKER

Dendrocopus minor

This is the smallest of our three resident woodpeckers and the least common. Broad-leaved woodland, mature hedgerows, parkland and shelterbelts are the preferred habitats, though individuals may visit garden feeding stations (much less frequently than our other two woodpecker species). Lesser Spotted Woodpeckers are very small birds (about the size of a House Sparrow) and can be separated from their larger cousins by the presence of a red crown and the lack of red under the tail.

Lesser Spotted Woodpecker

SAND MARTIN

Riparia riparia

Sand Martins are summer visitors, one of the first to arrive in spring, and become widespread across much of Britain and Ireland during the course of the summer. They nest in colonies of between 10 and 1,000 strong, excavating their burrows in soft sandy riverbanks or quarry faces. Feeding on the wing, often over water, Sand Martins take a range of insects, including larger flies and aphids. Although they do not use the supplementary food provided at garden feeding stations, they may be seen feeding overhead.

OUT OF THE ORDINARY

GREY WAGTAIL *Motacilla cinerea*

Although still regarded as a breeding bird of upland, fast-flowing rivers, the Grey Wagtail can now be found breeding on many lowland stretches of slow-moving water. Here it builds a nest on a ledge (often under a bridge) or among the roots of a riverbank tree. During the winter months, the upland breeding territories tend to be abandoned, the birds moving to warmer lowland sites where insects remain active. Included among these sites are gardens, with some individual birds returning to the same gardens in subsequent years to feed on insects, small seeds and fat.

Grey Wagtail

Black Redstart juvenile

RIGHT: **Firecrest male**

BLACK REDSTART

Phoenicurus ochruros

The Black Redstart is a scarce breeding bird, most often associated with industrial sites in highly urbanised landscapes. As well as breeding here, other individuals overwinter or pass through on migration. It is the migrant birds that are most likely to turn up in gardens, appearing briefly during the autumn or spring migration period.

Black Redstart adult male

RIGHT: **Firecrest male displaying.**

GARDEN WARBLER *Sylvia borin*

Despite its name, the summer-visiting Garden Warbler is really a bird of scrub and open woodland, where it breeds and feeds within the dense understorey. This species tends to be reported from gardens during the spring and autumn migration, periods when migrant birds often turn up in atypical habitats. During autumn migration these warblers make good use of available berry crops, increasing their resources prior to departure.

FIRECREST *Regulus ignicapilla*

This rare but increasing breeding bird is most commonly associated with coniferous and mixed woodland. However, like the related Goldcrest, it will utilise ornamental conifers and may sometimes be recorded from gardens, arriving alone or as part of a roving flock of tits and warblers during late autumn. Firecrests feed on small invertebrates, especially spiders, aphids and springtails, gleaned from twigs and branches in the canopy.

HAWFINCH *Coccothraustes coccothraustes*

Although found over much of England, Wales and parts of southern Scotland, the Hawfinch is a rarely seen bird. Highly secretive in nature and thinly distributed, it is one of the rarest of garden visitors. Outside of the breeding season, Hawfinches feed on seeds of Beech, Hornbeam and Wild Cherry. Once supplies of these seeds have been depleted, the birds may move into hedgerows and gardens in search of fruit and other seeds, before returning to the woods in early spring to feed on the newly developed buds.

male

male

BELOW: male

LINNET *Carduelis cannabina*

The Linnet is one a number of farmland birds whose populations have declined because of a change in farming practices. The species is largely dependent upon small seeds, typically those of annual weeds, throughout the year and as these decreased in their availability so the numbers of Linnets fell. In recent years numbers have stabilised, or even increased slightly, reflecting the increasing availability of seeds on set-aside land and exploitation of oilseed rape – a new food source for this bird. Linnets tend to breed in small, rather loose, colonies in scrub, gorse or mature hedgerows. They may visit gardens during the winter months, attracted by some of the new seed mixes.

male

LEFT: **male**

male

female

LEFT: **juvenile**

COMMON CROSSBILL

Loxia curvirostra

The Crossbill is a bird of coniferous woodland, breeding very early in the year and nesting in the upper branches of mature conifers. The crossed bill is ideally suited to prizing open conifer cones and the species feeds on conifer seeds and little else. Numbers may vary from one year to the next, reflecting the success of the breeding season and the number of individuals arriving from populations elsewhere in Europe. Gardens are not often used but birds may visit bird baths or garden ponds.

male

male

female

mixed flock

female

YELLOWHAMMER *Emberiza citrinella*

This farmland bunting is not uncommonly reported from rural gardens during the months of late winter and early spring. This is a time of the year when seed supplies within farmland may be scarce and the seed on offer at garden feeding stations becomes increasingly attractive. Numbers may reach double-figures at favoured feeding sites, the birds often associating with Reed Buntings, Tree Sparrows and various finches. The Yellowhammer is another of the farmland seed-eating species to have undergone a pronounced decline in abundance following changes in farming practice.

male

female

male

male

FURTHER READING

Burton, R. (2005). *Garden Bird Behaviour*. New Holland, London.
Couzens, D. (2004). *The Secret Lives of Garden Birds*. Christopher Helm, London.
Du Feu, C. (2003). *The BTO Nestbox Guide*. BTO, Thetford, Norfolk.
Flegg, J. (2004). *Time to Fly: Exploring Bird Migration*. BTO, Thetford, Norfolk.
Moss, S. (2004). *The Bird-Friendly Garden*. HarperCollins, London.
Snow, B. & Snow, D. (1998). *Birds and Berries*. T & A D Poyser, Calton.
Soper, T. (2006). *Tony Soper's Bird Table Book*. David & Charles, Newton Abbot, Devon.

Sterry, P.R. (2004). *Collins Complete British Birds*. HarperCollins, London.
Sterry, P.R. (2008). *Collins British Wildlife*. HarperCollins, London.
Svensson, L., Mullarney, K., Zetterström, D. & Grant, P. (2001). *Collins Bird Guide*. HarperCollins, London.
Toms, M. (2003). *The BTO/CJ Garden BirdWatch Book*. BTO, Thetford, Norfolk.
Toms, M., Wilson, I. & Wilson, B. (2008). *Gardening for Birdwatchers*. BTO, Thetford, Norfolk.

USEFUL WEBSITES

Botanical Society of the British Isles (BSBI) - www.bsbi.org.uk
British Trust for Ornithology (BTO) - www.bto.org
Countryside Council for Wales - www.ccw.gov.uk
The Forestry Commission - www.forestry.gov.uk
Herpetological Conservation Trust - www.herpconstrust.org.uk
Joint Nature Conservation Committee - www.jncc.gov.uk
The National Trust - www.nationaltrust.org.uk

The National Trust for Scotland - www.nts.org.uk
Natural England - www.naturalengland.org.uk
Plantlife International - www.plantlife.org.uk
Royal Society for the Protection of Birds (RSPB) - www.rspb.org.uk
Scottish Natural Heritage - www.snh.org.uk
The Wildlife Trusts - www.wildlifetrusts.org.uk
The Woodland Trust - www.woodland-trust.org.uk

PICTURE CREDITS

All photographs used in this book were taken by Paul Sterry with the exception of those listed below; these can be identified using a combination of page number and subject.

From the files of Nature Photographers Ltd
Dave Ashton: 107 Sparrowhawk eating Woodpigeon. S.C. Bisserot: 72 Great Crested Newt female; 87 Red-tailed Bumblebee; 87 Buff-tailed Bumblebee; 87 Common Carder Bumblebee. Frank Blackburn: 121 Little Owl at nest (left); 129 Green Woodpecker female at nest; 159 Blackcap female at nest; 171 Willow Tit at nest; 219 Lesser Spotted Woodpecker. T.D. Bonsall: 155 Mistle Thrush at nest. Idris Bowen: 72 Smooth Newt female. Nicholas Phelps Brown: 84 Cranefly; 89 Zebra Spider. Andy Callow: 47 Azure Damselfly; 85 Red Mason Bee; 85 Black Ant; 85 *Myrmica rufa*; 88 2-spot Ladybird; 89 Red Spider Mite. Kevin Carlson: 119 Collared Dove at nest; 161 Chiffchaff at nest. Colin Carver: 70 Roe Deer; 70 Fox; 70 Stoat; 70 Weasel; 71 Grass Snake; 131 Great Spotted Woodpecker male at nest; 171 Marsh Tit at nest; 183 Magpie at nest; 211 Reed Bunting male perched; 218 Stock Dove. Hugh Clark: 68 Mole; 71 Common Pipistrelle flying; 118 Tawny Owl at nest; 178 Treecreeper in flight; 197 Chaffinch female in flight; 221 Crossbill female. Andrew Cleave: 31 bird bath; 61 Ground-elder; 70 Badger. Ron Croucher: 17 Blue Tits; 121 Little Owl at nest (right); 167 Blue Tit in nestbox. Geoff du Feu: 83 Rose Aphid; 85 Greenbottle; 85 Common House-fly; 85 Bluebottle; 85 Leaf-cutter Bee. R.H. Fisher: 28 Blackbird on nest. Michael Foord: 184 Jackdaws on birdtable; 189 Rook on birdtable. Phil Green: 101 Nuthatch; 177 Nuthatch climbing down tree. Ernie Janes: 17 Chaffinch; 26 Great Tit family; 69 Common Rat; 70 Muntjac; 71 Brown Long-eared Bat; 75 Common Blue female; 89 House Spider; 105 Lesser Redpoll; 107 Sparrowhawk female at nest; 107 Sparrowhawk male perched; 111 Black-headed Gulls following plough; 113 Feral Pigeon flock; 145 Robin at nest; 169 Great Tit in nestbox; 203 Goldfinch perched (right); 209 Lesser Redpoll male on cone. Lee Morgan: 71 Grey Long-eared Bat; 71 Common Pipistrelle; 71 Soprano Pipistrelle. Owen Newman: 68 Hedgehog; 69 Yellow-necked Mouse; 71 Brown Long-eared Bat flying. Philip Newman: 153 Song Thrush at nest; 153 Song Thrush with elderberries; 175 Long-tailed Tit at nest; 209 Lesser Redpoll female drinking; 221 Crossbill male. W.S. Paton: 16 Blackbird at nest; 134 House Martin at nest; 146 Blackbird male at nest; 173 Coal Tit at nest. Richard Revels: 42 Great Tit family; 47 Common Darter; 47 Common Blue Damselfly; 47 Blue-tailed Damselfly; 69 Red Squirrel; 73 Small White; 73 Green-veined White; 73 Orange-tip; 73 Small Tortoiseshell; 74 Speckled Wood upperwing; 74 Ringlet upperwing; 74 Gatekeeper upperwing; 74 Meadow Brown upperwing; 75 Small Copper upperwing; 75 Common Blue male; 75 Holly Blue upperwing; 75 Small Skipper; 75 Essex Skipper; 75 Large Skipper; 77 Hummingbird Hawkmoth;

84 *Helophilus pendulus*; 86 German Wasp; 86 Common Wasp; 86 Median Wasp; 149 Fieldfare in flight; 165 Spotted Flycatcher with moth; 181 Jay with acorn; 191 Starling at birdtable. Don Smith: 119 Tawny Owl in flight; 179 Treecreeper roosting. E.K. Thompson: 118 Tawny Owl perched. Roger Tidman: 18 Starling flock; 104 Goldfinch juvenile; 112 Feral Pigeon at nest; 113 white dove; 123 Swift drinking; 133 Swallow in flight; 134 House Martin in flight; 136 Pied Wagtail male; 139 Waxwings on aerial; 143 Dunnock with Cuckoo; 147 Blackbird (leucistic bird); 147 Blackbird juvenile; 150 Redwing perched; 151 Redwing with berries; 172 Coal Tit in conifer; 186 Raven; 191 Starling flock; 202 Goldfinch juvenile; 212 Nightingale; 219 Sand Martin; 221 Hawfinch. Patrick Whalley: 215 Crested Tit. A. Wharton: 89 Crab Spider; 157 Whitethroat at nest.

From the files of the BTO Photolibrary
Dawn Balmer: 11 Birdwatchers; 160 Chiffchaff being ringed. John Black: 122 Swift, young at nest. Al Downie: 37 Birder with telescope. Mark Grantham: 10 BTO scientists. John Harding: 17 Bullfinches; 19 Kingfisher; 25 Waxwing eating berries; 28 Robin; 32 Brambling; 39 Siskin; 120 Collared Dove on post; 125 Kingfisher (bottom right); 128 Green Woodpecker young in nest; 135 House Martin with mud; 139 Waxwing eating berries; 141 Wren in winter; 143 Dunnock in winter; 145 Robin juvenile; 163 Goldcrest female at nest; 164 Spotted Flycatcher at nest; 195 Tree Sparrow bathing; 196 Chaffinch male calling; 197 Chaffinch male perched (right); 199 Brambling at birdtable; 205 Siskin at feeder; 207 Bullfinch female on ground; 210 Reed Bunting male at birdtable. Jill Pakenham: 19 Greenfinch and Blue Tit; 20 Blue Tit; 26 Blue Tit juvenile; 27 Magpie; 31 Whitethroat; 31 Greenfinches fighting; 107 Sparrowhawk bathing; 115 Woodpigeon on lawn; 121 Little Owl perched; 129 Green Woodpecker female on lawn; 130 Great Spotted Woodpecker male drilling wood; 132 Swallow, young begging; 132 Swallow at nest; 140 Wren, Shetland race; 157 Whitethroat bathing; 166 Blue Tit in snow; 167 Blue Tit perched on willow; 167 Blue Tit juvenile; 173 Coal Tit perched; 175 Long-tailed Tit perched; 177 Nuthatch on feeder; 184 Jackdaw on grass; 201 Greenfinch fighting; 203 Goldfinch flying; 207 Bullfinch male perched (right). Paul Stancliffe: 36 Garden birdwatchers. Mike Toms: 10 BTO headquarters; 10 BTO fieldwork; 11 Pied Wagtail being ringed; 11 Reed Warbler in mist net; 16 Collared Dove; 19 Robin sunbathing; 27 Cat; 29 Garden; 46 Garden pond; 115 Woodpigeon on wall; 153 Smashed snail shell.

Artists whose work appears in this book are as follows: Richard Allen, Norman Arlott, Trevor Boyser, Hilary Burn, John Cox, Dave Daly, John Gale, Robert Gillmor, Peter Hayman, Ian Lewington, David Quinn, Darren Rees, Chris Rose, Christopher Schmidt.

INDEX